DANGEROUS MARINE ANIMALS

DANGEROUS MARINE ANIMALS

THAT BITE, STING, SHOCK, ARE NON-EDIBLE

Second Edition

BRUCE W. HALSTEAD, M.D.

Director, International
Biotoxicological Center
World Life Research Institute
Colton, California

CORNELL MARITIME PRESS
Centreville, Maryland

Library of Congress Cataloging in Publication Data

Halstead, Bruce W
Dangerous marine animals.

Bibliography: p.
Includes index.
1. Dangerous marine animals. I. Title.
[DNLM: 1. Animals, Poisonous. 2. Marine biology.
WD400 H196d]
QL100.H3 1980 591.6'9 80-15475
ISBN 0-87033-268-6

Manufactured in the United States of America

First edition, 1959; Second edition, 1980

To my wife, Joy,
and children, Linda,
Sandy, David, Larry,
Claudia, and Shari

CONTENTS

PLATES

(Plates 1-8 appear following page 52)

(Plates 9-16 appear following page 148)

PREFACE

The original volume of *Dangerous Marine Animals* was published in 1959. Since that time a great deal of new material has become available in widely scattered places which needs to be drawn together in a single, up-to-date manual. Furthermore, with the enormous expansion of human activities in oceanology and aquatic sports, encounters with noxious marine organisms have increased significantly in recent decades. For these reasons it seemed timely to undertake the preparation of a revised edition of this work.

The purpose of this new edition is to provide a ready source of information regarding the indentification, geographical distribution, habits, and noxious characteristics of dangerous marine animals. Sufficient data have been included to be of value to someone administering first-aid as well as to the physician. Technical discussions of the chemistry and pharmacology of these poisonous substances involved, most of which are unknown, have been purposely eliminated. Every effort has been taken to keep this a practical guide, keeping technical details to a minimum. If this volume can contribute in a small way to the health and happiness of those working with marine organisms in their native habitats, it will have fulfilled its mission.

Much of the data presented in the revised edition of this book is the result of numerous gifts, grants, and contracts among which include the following agencies: School of Aerospace Medicine, Air Force Office of Scientific Research, and Science Division, Directorate of Science and Technology, Department of the Air Force; Office of Naval Research, Bureau of Medicine and Surgery, Department of the Navy; Chemical and Radiological Laboratories, Office of the Surgeon General, Department of the Navy; National Institutes of Health, U.S. Public Health Service; National Science Foundation; Armed Forces Institute of Pathology; U.S. Coast Guard; Pacific Oceanic Fishery Investigations, U.S.

Fish and Wildlife Service; Van Camp Laboratories; Food and Agriculture Organization, United Nations; American Philosophical Society; William Waterman Fund of the Research Corporation of America; National Library of Medicine; Pacific Science Board, National Research Council; U.S. National Museum, Smithsonian Institution.

For their many and varied contributions, I am deeply grateful to all of these organizations and their staffs.

I am also deeply indebted to the following artists whose works so materially enhance this volume: Shirgeru Arita (Pl. 13-C, 15-D; Fig. 70-B, 98); Harry Baerg (Pl. 6; Fig. 11, 18, 103, 104, 105); Lou Barlow (Pl. 7-A); Sharon K. Calloway (Fig. 84); George Coates (Pl. 10-E; Fig. 79-B); A.M. Cousteau (Fig. 10-A); K. Cox (Fig. 93-B); Charles Cutress (Fig. 26-A); John Garth (Fig. 93-C); Padres Giacomelli (Pl. 10-A); Keith Gillett (Pl. 1, 5; Fig. 26-B, 30-B); M. Holmes (Fig. 48); O. Hon (Fig. 19, 76); L.C. Innes (Fig. 64); T. Iwago (Fig. 42); Robert Harold Knabenbauer (Pl. 2, 3, 9; Fig. 8-B, 10-B, 15, 17, 33, 95); Robert Kreuzinger (Fig. 29, 35, 36, 37, 38, 43, 62, 66-D, 72, 75, 80, 82, 87, 88, 89, 90, 91, 108, 110); Toshio Kumada (Pl. 11-C, 14-A, 15-B; Fig. 9-C, 14); George Lower (Pl. 10-B); D. Masry (Fig. 28-A); D. Ollis (Fig. 12); A. Powers (Fig. 41); R.C. Schoening (Pl. 4, 8); Mitsuo Shirao (Pl. 2-A; Fig. 6, 7, 8-A, -C, -D, 9-A, -B, 52, 53-B, 65-B, 67-B, 68-C, 69-C, 70-A, 74-B, 77, 79-A, 85-D, 93-A); J. Siebenaler (Fig. 54); Robert Straughan (Pl. 10-C); John Tashijan (Fig. 109); K. Tomita (Pl. 11-C, 16-B, -C); and Talbot H. Waterman (Fig. 92).

— Bruce W. Halstead

Colton, California
June, 1980

DANGEROUS MARINE ANIMALS

I

DANGEROUS MARINE ANIMALS— OUR KNOWLEDGE OF THE PAST

The development of the aqualung by the great French diver Jacques Yves Cousteau and his colleagues gave birth to a new area of activity for mankind—skin diving. A new environment was discovered, called by Cousteau the "silent world." Indeed it is a seemingly silent world, enshrouded in mystery, intrigue, and beauty beyond description. Tropical reefs offer a submarine panorama of colors that seem to vibrate and sparkle. But with this spectacular beauty there is often great peril. In multiplied instances, beauty of form, gracefulness of carriage, lavishness of color seem to go hand in hand with disease and death. Hidden within the delicate lacy fins of the magnificent zebrafish are the needle-like hollow spines that convey deadly venom. Dangers inherent in certain types of marine life are sufficiently great to behoove one to develop an intelligent appreciation of them.

From time immemorial people have depended upon the sea for food, medicine and clothing and, in recent times, for many industrial and military needs. Philosophers as well as scientists have peered into the depths of the sea and speculated about its contents. What man has been unable to observe directly, his imagination has fashioned in myth and fable. The abysses of the sea and its little-known animal life have served well the interests of writer and story-teller. However, as has been repeatedly shown, much folklore about the sea is based on elements of fact. Therefore, it is of some interest to explore briefly the evolution of our knowledge of dangerous marine organisms.

SHARKS

Sharks and their attacks upon man have stimulated the imagination of writers from antiquity. Probably the earliest reference to man's encounters with sharks is by Pliny the Elder who wrote during the first century of the "cruel combat that sponge fishermen must maintain against the dogfish. . . ." He pointed out that dogfish attack the groins, the heels and all the white portions of the body. He recommended the best way to frighten off the dogfish was to swim straight for it. Pliny also believed the danger to a diver from shark attacks was greater near the surface of the water than at deeper levels. It is interesting to note that many of Pliny's observations are accepted facts in some diving circles today.

Guillaume Rondelet, famous French physician and naturalist of the Rennaissance, frequently lacked facts, but never originality. He describes a footman who was reportedly "observed" running down the beach pursued by a dogfish hot at his heels. Fortunately, the footman struck at the beast with his foot and killed it. In those days dogfish must have been marathon runners! It was Rondelet who also contended that Jonah of the Scriptures was swallowed by a shark rather than a whale.

One of the earliest references to a shark actually attacking a man is in *Septentrionalibus* written and published about 1555 in Rome by Olaus Magnus. Included in his work is a drawing of several sharks attacking a bather while the bather is being rescued by a "kindly" ray. In 1623, the Dutch navigator Carstenszoon recorded finding sharks, swordfish and other "unnatural monsters" from the waters near Cape York, Queensland, and alluded to some of their dangers.

According to Dr. Paul Budker, a modern French scholar, sharks were regarded with great fear by sailors working in the Mediterranean during the 18th Century. Believing that a shark would not attack except in hunger, the sailors felt compelled to throw a loaf of bread to the monster. If this did not suffice, a sailor would have to be lowered by a rope to the surface of the water and look menacingly at the shark—otherwise it was believed, the shark would grab the vessel with its teeth and devour it.

Probably the greatest number of shark attacks have taken place in Australian waters. The first recorded shark tragedy in that region is by Watkin Tench in *The Narrative of the Expedition to Botany Bay* (1789). Tench refers to a female aboriginal who was bitten in half by a shark in New South Wales several years before. Francois Peron, the naturalist for the Baudin expedition, described a shark attack which occurred at Faure Island, Hamelin Harbour, Western Australia, in 1803.

The question of whether sharks prefer dark or light-skinned victims has been a favorite topic for argument among old "salts." The question appears to have originated with Count Bernard Lacépède, the famous

French naturalist, who vigorously debated the subject with some of his colleagues during the early 1800's. Lacépède was of the firm opinion that sharks were attracted to negroes because of their "more odorous emanations." However, modern research has shown that sharks seem to be more readily attracted to light-colored objects.

Paul Budker notes in his *La Vie des Requins* that the old slave ships did much to encourage the anthropophagic tastes of sharks. Negro slaves were transported in ships in the most primitive, unhygienic and revolting conditions imaginable. Disease and death were rampant under these circumstances. Trailing in the wake of these vessels of "man's inhumanity to man" were schools of sharks which seemed to await with ghoulish delight their daily ration of human corpses tossed over the side. After one severe storm on a run between Mozambique and Bourbon, France, more than fifty slaves suffocated because of the necessity of closing the hatches. The log of the vessel read: "What a windfall for the sharks!"

Although man's involvement with sharks and shark attacks dates back to ancient times, most authoritative data on the subject have appeared in technical publications only within the past two decades. The work by Dr. Gilbert P. Whitley (1940), *The fishes of Australia. The sharks, rays, devil-fish, and other primitive fishes of Australia and New Zealand*, contains a great deal of valuable and interesting information regarding the habits, identification, and folklore connected with sharks. His work also includes a list of all the reported shark attacks in Australia to the time of publication in 1940.

The increased loss of surface vessels and aircraft, and the resulting high incidence of shark attacks during World War II, pointed up the need for an effective shark repellent. Through the combined efforts of the U.S. Office of Scientific Research and Development, U.S. Navy and Army Air Corps, the National Research Council, and various civilian organizations, such as the American Museum of Natural History, Woods Hole Oceanographic Institution, Scripps Institute of Oceanography, Marine Studios, Inc., New York Zoological Society, University of Florida, and the Calco Chemical Corporation, an intensive search was made for a satisfactory deterrent. Experiments were conducted feeding sharks bait with the various chemicals to be tested. Controls were also run feeding the sharks bait without the repellent, and the results of the two tests were then compared. After testing more than 70 different chemical compounds, it was found that no more than four were deserving of further attention. These were: decomposing shark meat, maleic acid, copper sulphate, and copper acetate. It was subsequently shown that the potent factor in the decomposing meat was ammonium acetate. It was later decided that a dye should be added in order to hide the swimmer from view. Finally a chemical packet was developed which combined a nigrosine dye with copper acetate. This compound was later given the popular but somewhat misleading name, *shark chaser*. According to the research reports conducted at that time, this compound was reputed to be highly effective under normal feeding conditions of the shark. More recent studies have shown that the repellent is generally ineffective under circumstances when it is needed most. Nevertheless, it was the best repellent available for many years and was used extensively on life preservers for the U.S. Armed Forces. The material was commercially produced by the Shark Chaser Chemical Company, San Pedro, California.

About the time that American researchers were conducting their investigations on shark repellents, similar studies were going on in Australia under the auspices of the Commonwealth Scientific and Industrial Research Organization. Because of the high incidence of shark attacks along the coast of New South Wales, Australia, the local government instituted a Shark Menace Advisory Committee to investigate methods of protecting bathers from shark attacks. As a result of their studies, a program in meshing was begun, consiting of laying a long rope net overnight near the beach and removing it by trawler in the morning. This technique has proved to be highly efficacious and is still employed in Australia, South Africa, and in other areas where sharks are considered a menace to bathers.

One of the earliest and best documented serious publications to deal with the overall problem of shark attacks appeared in 1958 by Victor M. Coppleson, the noted Australian physician, who died in 1963. His well illustrated book, entitled *Shark Attack*, contains numerous accounts of attacks, resultant medical injuries, methods of dealing with an attacking shark as well as identification, habits, and distribution of dangerous shark species. Coppleson's work is a classic in its field and should be read by anyone interested in the subject.

Probably the greatest impetus to the scientific investigation of shark attacks and related medical problems was the result of the American Institute of Biological Sciences' Conference on Shark Repellents held in New Orleans (April, 1958), sponsored jointly by the AIBS,

Tulane University and the Office of Naval Research and the Bureau of Aeronautics of the U.S. Navy. Although the main topic of the conference was concerned with the efficacy of the "shark chaser" which had been developed by the U.S. Naval Research Laboratory during World War II, it soon became evident to everyone that there was a vast gap in our basic knowledge of the biology, habits, and even the identification of dangerous sharks. Out of this conference came a series of recommendations concerning the need for further information on shark behavior, ecology, taxonomy, physiology, morphology, and particularly the teleceptor sensor mechanisms of sharks. An international committee known as the Shark Research Panel was appointed to formulate tentative recommendations for a broad scientific program dealing with the shark hazard problem under the capable chairmanship of Perry A. Gilbert, and originally consisted of Lester R. Aronson, Bruce W. Halstead, Carl L. Hubbs, Arthur W. Martin, Leonard P. Schultz, J. L. B. Smith, Stewart Springer, Tohru Uchida, F. G. Wood, Jr., and Gilbert P. Whitley. A group of 34 scientists internationally recognized for their knowledge of sharks participated in the original conference. One of the significant outgrowths of this conference was the ultimate establishment of the worldwide Shark Attack File. Initially, the Shark Attack File was housed at the Smithsonian Institution under the direct supervision of Leonard P. Schultz. Later it was moved to the Mote Marine Laboratory in Sarasota, Florida. Gilbert served as chairman of the Shark Research Panel during the almost 12 years of its existence. The Shark Attack File contained data on a series of 1,165 case histories covering the years 1941 through 1968, and was ultimately expanded to include 1,652 cases. Final analysis of the Shark Attack File was done by a group of statisticians and programmers of the U.S. Navy's Bureau of Medicine and Surgery, and the final publication appeared as a Special Technical Report (Contract No. N00014-73-C-0252), dated 31 October 1973 under the authorship of H. David Baldridge, entitled *Shark Attack Against Man*. A popularized, paperback version of Baldridge's work, entitled *Shark Attack*, was published later in 1974. A distillate of some of the more pertinent findings obtained from the Shark Attack File appear in the section of this handbook dealing with sharks.

With the expansion of underwater recreational activity and the maintenance of large oceanariums, our knowledge of other traumagenic marine organisms has increased significantly. Some of this additional knowledge is reflected in this latest revision.

STINGING MARINE ANIMALS

Our knowledge of stinging marine animals dates back to the days of classical antiquity. Aristotle, writing from 348 to 322 B.C., made reference to the stinging ability of stingrays. Nicander, Greek poet and physician of the second century B.C., reported that the sting of the stingray caused "gangrene of the wounded flesh." Pliny the Elder, carried away with the subject, reported that the deadly stingray, called Trygon, was able to kill a tree by driving its sting into one of the roots! He also mentions the serious wounds produced by the weeverfish.

Numerous scientific and popular reports are to be found in early literature relative to the dangers of stings from venomous fishes, but serious scientific research on the subject was not conducted until the time of Dr. A. Bottard, a French scientist, who wrote the first monograph on the venom organs of venomous fishes, entitled *"Les Poissons Venimeux"* and published in 1889. Several scientists previously worked on the venom organs of weeverfishes, but Bottard was the first to approach the subject in a more or less organized manner. Another important contribution to the field of marine venomology was that of the great French physician Albert Calmette, whose classic *Venoms, Venomous Animals, and Antivenomous Serum-Therapeutics* was published in 1908.

One of the earliest known occupational illnesses of divers caused by the biological hazard of stinging marine animals is the so-called sponge fisherman's disease. This disease, frequently manifested by a severe burning skin rash with areas of painful ulcerations, for many years was believed to be due to the handling of sponges. Finally, Dr. S. G. Zervos demonstrated in a series of scientific articles published in France in 1903, 1934, and 1938 that the disease was not due directly to sponges, but rather to coming into contact with the stinging tentacles of small sea anemones adhering to the sponges.

We are largely indebted to Dr. H. Muir Evans, an English surgeon, for our knowledge of the venom apparatus of the stingray. Much of Dr. Evans' medical practice was concerned with the care of fishermen in Lowestoft, England, who came to him with wounds they had received from weevers, spiny dogfish, and stingrays. The stimulus Dr. Evans received from his patients goaded him to study the venomous nature of these fish on a scientific basis. Thus a new horizon opened in medicine, described by Evans in an interesting manner in his *Stingfish and Seafarer* (1943). Those who have engaged in research on venomous fishes in

America owe much to the challenge provided by the writings of this fine old English surgeon of Lowestoft. Serious students interested in stinging marine organisms will also receive much benefit from the outstanding volumes of Marie Phisalix, "*Animaux Venimeux et Venins*" (1922), and the brilliant writings of E. N. Pawlowsky, "*Gifttiere und Giftigkeit*" (1927).

Modern marine venomologists are indebted to the numerous contributions of Australian scientists who have studied the identification, morphology, and toxicology of some of the oceans' most dangerous stinging creatures inhabiting the southwestern Indo-Pacific region. Much of this research is discussed in the well-illustrated volume by John B. Cleland and R. V. Southcott, entitled *Injuries to Man from Marine Invertebrates in the Australian Region.* Carl Edmonds has produced a valuable handbook on the *Dangerous Marine Animals of the Indo-Pacific Region.* Noteworthy also is the valuable Japanese work *Marine Toxins*, written by one of the world's foremost marine biotoxicologists, Yoshiro Hashimoto. This book, the last contribution of a great scientist, was completed just prior to his death. At the moment the volume is limited to the Japanese language. It is hoped an English edition will be published.

Sea snakes are a large group of venomous marine organisms about which very little is known concerning their venoms, venom apparatus, or biology. It is therefore with a great deal of interest that marine venomologists view the work *The Biology of Sea Snakes*, edited by William A. Dunson and published by the University Park Press, Baltimore, in 1975. This revised edition of *Dangerous Marine Animals* includes some of the more recent knowledge concerning the nature of the venom apparatus and venoms of these fascinating aquatic reptiles.

MARINE ANIMALS POSIONOUS TO EAT

The danger of eating poisonous marine organisms was known in the time of the Pharaohs of ancient Egypt, during the Toltec civilization in ancient Mexico, and in Biblical days when Moses led the children of Israel out of the land of Egypt with the warning "... and whatsoever hath not fins and scales ye may not eat; it is unclean unto you" (Deut. 14:9-10). This advice is still recommended in modern American military survival manuals.

Emperors of ancient Japan forbade soldiers to indulge in the most poisonous of all fishes, the deadly puffer, called in Japan, the fugu. Those caught eating this scaleless delicate morsel lost their entire inheritance. Similar prohibitions were issued by Alexander the Great to his soldiers.

Peter Martyr, the great historian of the West Indies, specifically refers to the danger of the ciguatera type of fish poisoning. This form of poisoning supposedly developed in fishes as a result of their eating the poison machineel berry—a theory which continues to persist today, but is without scientific foundation. As a result of eating poisonous fishes sold to him by friendly savages in New Caledonia, Captain Cook almost terminated his famous world voyage in 1776.

Through the centuries, hundreds of articles written by explorers, missionaries, physicians, naturalists, and scientists describe the dangers of eating molluscs, fishes, and other marine organisms at certain seasons of the year under given circumstances. The French, Russian, Japanese, and American navies have called attention to the medical importance of these organisms. It is estimated that during World War II more than 400 Japanese military personnel lost their lives in Micronesia from the eating of fresh tropical reef fishes. A most unusual piece of toxicological research on the poisonous fishes of Micronesia was conducted by the Japanese scientist, Dr. Yoshio Hiyama. Visiting the Marshall and Saipan Islands under the most trying of circumstances, he found that large numbers of the reef fishes of these islands were deadly for humans to eat. It was Dr. Hiyama's work that served as a basis for much of the subsequent research on poisonous fishes conducted in the United States and elsewhere.

The origin of oral fish intoxicants, fish that are poisonous to eat, has always been a mystery, and continues to elude scientists. For centuries it has been known that a fish species may be edible in one geographical area and deadly poisonous in another. There are a number of different types of oral fish poisons. Some of them are extremely rare, whereas others are of common occurrence in tropical insular areas of the world. Scombroid fish poisoning, ciguatera, and puffer poisoning continue to remain significant public health problems in many tropical islands, and threaten the development and utilization of shore fisheries in areas wherever these public health problems are endemic.

During the last two decades a great deal of scientific effort has been directed toward obtaining an understanding of the chemical and pharmacological properties of ciguatoxin, saxitoxin, and tetrodotoxin (the cause of puffer poisoning). The chemical structure of saxitoxin and tetrodotoxin is now well-known, and probably more than 500 scientific articles concerning their pharmacological and neurophysiological properties have been written. However, the biogenesis of ciguatoxin continues to remain unknown although recently it has been demonstrated that dinoflagellates are probably involved. Despite the relatively common occurrence of scombrotoxin, the cause of scombroid fish poisoning, little is known concerning its precise chemical structure.

The present handbook deals with only a smattering of facts regarding animals of marine origin that are poisonous to eat. However, an attempt is made to present those facts of crucial significance to the aquatic enthusiast. One of the first things to remember is that not everything in the sea is edible, and that one should be cautious about eating inshore fishes in tropical island areas without first checking with the local inhabitants. Otherwise, that fish dinner about to be eaten might be the last one! If mankind is to continue to utilize the resources of the sea, greater efforts must be expended to attempt to unravel the mysteries hidden in its depths.

WHAT OF THE FUTURE?

As we glance at the dimly lit pages of the past, we are reminded of the debt of gratitude the scientific community owes to scientists of yesterday. Old musty archives continue to contribute valuable medical data to those who go into the depths of the sea. Our applied knowledge regarding most marine creatures is still very meager. Moreover, our knowledge will continue to remain so until basic medical and marine biological research is permitted to join hands in pushing back the frontiers of ignorance.

The reader should keep in mind that as we enter the sea and match wits with its death-dealing creatures, there is an ever-increasing amount of scientific evidence that these same noxious organisms and their poisons may serve as sources of the life-giving antibiotics, anticancer, and other indispensable therapeutic agents of tomorrow. These organisms must be considered a more than mere marine biological hazards—they are the vast chemical reserves of the future.

But now for the cold realities of today—dangerous marine organisms—their noxious effects, treatment, or better yet, how to avoid them!

II

MARINE ANIMALS THAT BITE

SHARKS

Sharks are the marine animals generally feared by most swimmers and divers. With the plethora of publications and movies dealing with shark attacks, real and fictitious, public awareness of the potential danger of sharks has reached an unprecedented high. During the past two decades a great deal of valuable information on the biology of sharks has come to light. However, the scientific community continues to suffer from a general lack of knowledge concerning these magnificent, toothy creatures. The following is a summary of some of the more pertinent information concerning sharks and their predatory habits. Much of this data is based on the International Shark Attack File investigations.

General: There are about two hundred fifty species of sharks known to ichthyologists, but only 27 species appear to have been definitely incriminated in attacks on man. Dangerous sharks are almost cosmopolitan in their geographical distribution (Figure 1). However, most shark attacks have occurred between latitudes 47° south and 46° north. The most northerly unprovoked shark attack occurred in the upper Adriatic Sea, and the most southerly attack took place off South Island, New Zealand. About 54 percent of the documented attacks have taken place in the Southern Hemisphere. None have been reported in the Arctic or Antarctic regions. Shark attacks are most common in warm temperate and tropical waters, occurring in areas where the water temperature is above 68°F (20°C).

Shark attacks tend to be seasonal depending upon weather conditions, warm water temperatures, and time of day when swimmers are most plentiful. Studies show that there is greater danger during late afternoon and at night, when sharks are most actively feeding.

Sharks range in size from some of the smaller species, which mature at a length of only 6 inches (15 cm), to the giant whale shark, which may attain a length of more than 50 feet (15 meters). Fortunately for the human species, the giant whale shark feeds primarily on planktonic organisms and its dentition is not suitable for larger objects.

The natural food of the larger, dangerous sharks consists mainly of large fishes, other sharks, seals and a variety of marine organisms. If compelled to generalize about the dietary habits of these animals, one may assume that if the shark is more than 4 feet (1.2 meters) in length, has adequate teeth, and if there is food or blood in the water, it is potentially dangerous to any human being nearby.

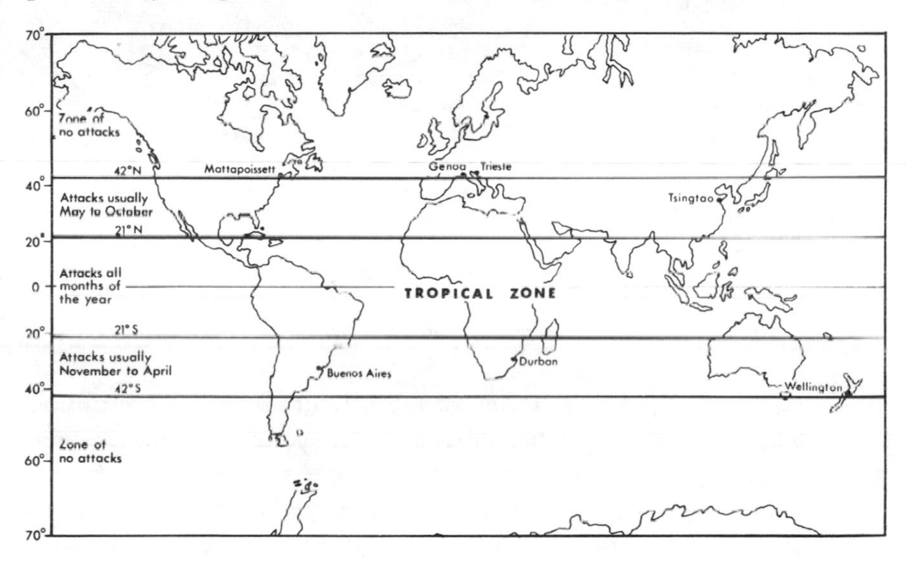

Figure 1. Map showing shark attack zones. (After Coppleson)

The sex of a shark can be determined by the presence or absence of claspers, appearing as modified pelvic fins in the male to form intromittent organs. The claspers are used by the male to convey the semen to the female. All sharks practice internal fertilization. There are three different ways by which sharks are born: from eggs (oviparous); alive (viviparous); or from eggs hatched within the uterus and then born alive (ovoviviparous). The method of birth employed varies with the species of the shark. The general external anatomy of a shark is shown in Figure 2.

Dentition: Sharks are characterized by having several rows or series of teeth in each jaw. The number of series of teeth in actual use at any one time varies from one to five in different species of sharks. Examination of the inside of the jaws will reveal that there are one to several additional reserve series lying in a reversed position. The reserve teeth point up in the upper jaw and down in the lower jaw. As functional teeth are lost, those of the next series move forward to replace them (Figure 3). This process of tooth replacement continues throughout the life of the shark. Teeth may be replaced singly in some shark species, or as an

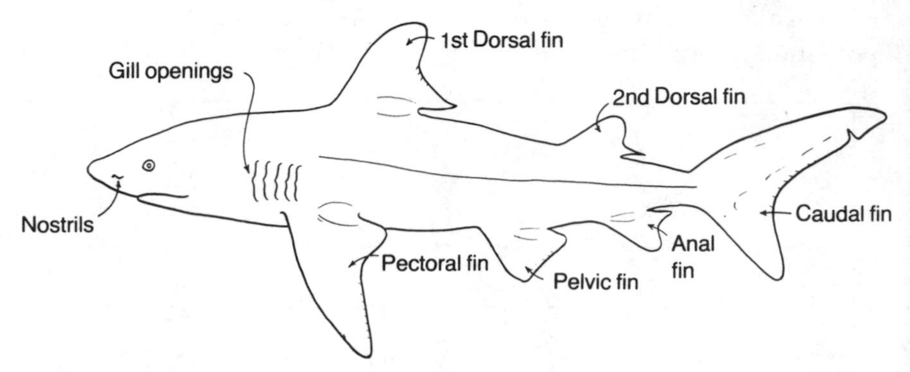

Figure 2. External anatomy of a typical shark.

entire series in others. A series of teeth which forms a continuous cutting edge is generally replaced as an entire series. The replacement process usually requires several days to a week or more.

The force of a shark bite is enormous. Actual measurements made with the use of a shark bite meter have shown that some of the larger sharks, such as the dusky shark (*Carcharhinus obscurus*), can exert a force of about 18 metric tons per square inch of tooth surface (30 kilograms per square millimeter).

Sensory System: The intelligence of a shark is generally considered to be of a low order, but this is of little consolation to anyone threatened by an attack. The means employed by sharks to detect their prey, food, or injured animals is a subject of prime importance to anyone concerned with shark attacks. Sharks have the usual sensory organs consisting of eyes, an auditory apparatus, and nose. The visual acuity of sharks appears to be better than previously believed, both for color and for distinguishing an object that is moving past a dark background. Vision is believed to be the prime sensory organ employed at a distance of 50 feet (15 meters) or less depending upon the clarity of the water. At distances greater than 50 feet, olfaction and other sensory organs appear to be of greater importance in food detection. The musculature of the eye permits the shark to maintain a constant visual field even when it is twisting, turning, or moving ahead. Since sharks spend much of their time at depths where lighting conditions are poor, some shark species have a special mirrorlike reflecting layer (*tapetum lucidum*) that is situated under the retina and reflects incoming light back through the

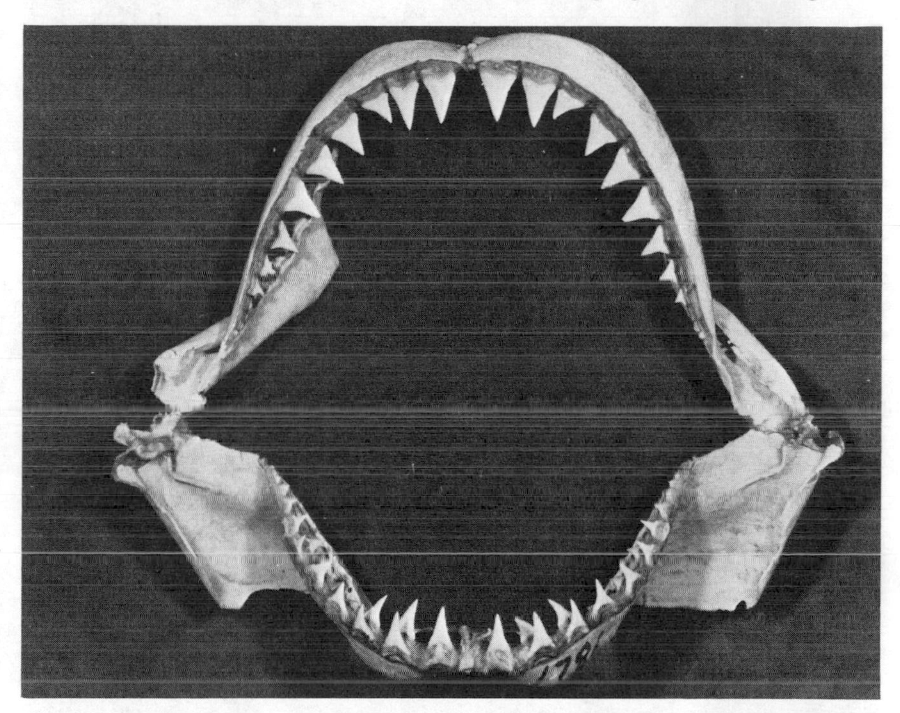

Figure 3. Jaws of the **Great White Shark,** *Carcharodon carcharias.* Anterior view. Length of shark about 8 feet (2.4 meters). (Courtesy Oceanographic Research Institute, Durban, South Africa)

retina. This mechanism permits optimum utilization of light in a dimly lit aquatic environment. Thus, the eyes of a shark are highly effective for its predatory activities.

Sharks have an amazingly acute sense of smell. As every comparative anatomist knows, the olfactory lobes of sharks are highly developed and comprise a major segment of their brain. Olfaction enables a shark

Figure 4. A. **Bull Shark,** *Carcharhinus leucas.* Young specimen. About 5 feet (1.5 meters) in length. B. **Dusky Shark,** *Carcharhinus obscurus.* About 5 feet (1.5 meters) in length. Sharks are in a "normal" swimming pattern. (Courtesy Oceanographic Research Institute, Durban, South Africa)

to detect blood and food in the water—probably at levels as minute as parts per billion or even more.

Sharks also have gustatory chemoreceptors or taste buds providing them with a sense of taste and permitting them to discriminate between food and other objects. These gustatory receptors undoubtedly play an important role in determining the dietary activities of sharks.

Another important set of sensory organs in sharks are the chemoreceptors in the skin, sometimes referred to as the "common chemical sense." These skin chemoreceptors provide the shark with the ability to perceive substances of an irritating nature, changes in salinity, water movements, and other chemical changes in the water. Knowledge concerning the precise function of these chemoreceptors could be of importance for the future development of chemical shark repellents.

One of the most important teleceptor devices in the shark is its "hearing" mechanism. Sharks detect sounds by means of the labyrinth (inner ear), the lateral line, and a group of specialized sensory receptors known as the *ampullae of Lorenzini*. With the use of these highly sensitive teleceptors, sharks have a phenomenal ability to detect low frequency vibrations in the water. This is why sharks can readily detect at great distances such disturbances as those caused by fish struggling and people swimming. The movements of sharks are controlled to a large extent by means of their combined teleceptor systems. The swimmer should keep in mind that sharks are always "listening" and are sensitive to any vibrational disturbance in the water.

When all of the various sensory systems of sharks are taken into consideration, it becomes obvious that the shark is a highly integrated "computerized" sensory marvel, possibly without equal either in the aquatic environment or on land.

Feeding Patterns: The feeding habits or patterns of sharks are quite variable but for convenience sake can be grouped into two general types: 1) *Normal feeding pattern*: This pattern can best be described as a situation in which one or several sharks are seen moving about with seemingly purposeful movements either in quest of food or actually feeding. In many instances the movements of the shark appear to be slow and determined, but at times may be erratic and swift moving. The swimming pattern, approach, and the final attack may vary with the species and circumstances (Figure 4). 2) *Frenzied feeding pattern*: The frenzied feeding pattern most frequently develops after a catastrophic event such as an explosion, sinking of a vessel, the crash of a plane in the water, or the sudden availability of large quantities of food. When any of these conditions is present, the entire feeding characteristics of sharks may rapidly change. The frenzied feeding pattern is further enhanced if sharks begin to congregate in large numbers (Figure 5). Under these conditions the swimming behavior of sharks becomes exceedingly erratic. They have been observed swimming vertically up to the surface, snapping at floating objects, and then suddenly plunging to the bottom of a lagoon, literally banging their noses on the bottom, and snapping savagely at anything in sight. It is during periods such as this that most repellent devices become useless and may even aggravate the situation still further. Cannibalism was observed in one instance during frenzied conditions, stimulated by an underwater blast, in which a school of snappers (*Lutjanidae*) had been killed. While one large shark fed on a snapper that had been caught under a coral head, a second shark moved into the scene and in one quick gulp ripped out the belly of the first shark. Within a few moments the hapless shark was completely devoured by other members of the pack who joined the killing. The species involved was the gray reef shark (*Carcharhinus amblyrhynchos*) at Jaluit, Marshall Islands.

Attack Pattern: In most instances (about 63 percent) the attacking shark was not seen by the victim prior to contact. In some instances sharks have been observed swimming very erratically just prior to making a rush on the victim. At the same time the pectoral fins were extended downward and the back of the shark was bowed into a hunched position. Then the shark was seen to swim stiffly with its whole body, its head moving back and forth almost as much as its tail. This behavior preceded a swift pass at the victim, a type of activity designated as "agonistic or combative behavior."

There is no single behavior performance which typifies the attack pattern of all sharks. Almost every conceivable type of activity has been observed under the heading of aggressive behavior. Shark attacks on man are sometimes preceded by one or more contacts which range from gentle bumping to violent collisions. At times these violent impacts can be severe enough to knock the victim completely out of the water.

Sharks have sometimes been known to single out an individual in a crowd of swimmers and continue to attack that individual to the exclusion of all other persons, including those who attempt to rescue the victim. However, because of the unpredictability of sharks, it is difficult to anticipate any specific type of attack approach.

Provoked versus Unprovoked Attacks: A significant number of shark attacks have occurred because of *provoked* encounters, such as spearing, poking at the shark, grabbing the shark by the tail, tantalizing the shark with fish, blocking off an escape passage way, or by some other means that obviously annoyed the shark. Encounters of this type are very foolhardy and should be avoided. Nevertheless, *unprovoked* attacks have taken place in which the victim appeared to be minding his

Figure 5. **Dusky Sharks,** *Carcharhinus obscurus.* Young specimens. About 5 feet (1.5 meters) in length, in a frenzied feeding pattern. Chunks of fish are in the water. Sharks in this frenzied behavior become very dangerous and erratic. (Courtesy Oceanographic Research Institute, Durban, South Africa)

own business and there appeared to be little or no reason for the attack—other than a hungry shark and the presence of a dainty morsel.

Color or Pattern Attraction: There are no firm, general conclusions that can be given about color and patterns relative to their attraction to sharks. However, it has been observed that bright highly reflective dyes, pigments, bright metallic objects, and contrasting color patterns may attract sharks. On the other hand, one should attempt to avoid appearing as a marine mammal or some other organism on which sharks normally feed. Obviously, what attracts a shark under one circumstance is possibly a deterrent in another set of environmental conditions. It should be kept in mind that the visual acuity of a shark may not always be the best under adverse environmental conditions. Consequently, when operating under conditions of poor visibility such as murky water or restricted lighting—color and pattern may be matters of increased importance in attracting sharks and precipitating an attack.

Incidence of Shark Attacks: Evaluation of the Shark Attack File revealed a documented incidence of about twenty-eight attacks per year, but there is reason to believe the overall incidence is much higher. The reason for this apparent discrepancy is that many attacks occur in remote areas of the world and pass by unreported. However, it is doubtful the actual incidence attains the 100 attacks per year that is sometimes given.

The overall case fatality rate based on documented records of a series of 1,165 attacks was 35 percent. Previous estimates were placed at a much higher figure.

SPECIES OF SHARKS DANGEROUS TO HUMAN BEINGS

The following is a partial listing of sharks known to be dangerous which have been incriminated in attacks on human beings. There are additional species known to be dangerous, but they have not been included because attacks by them are relatively uncommon. The families are arranged in alphabetical order.

Family *CARCHARHINIDAE*

Gray Sharks, Whaler Sharks, Requiem Sharks. This is the largest family of living sharks, comprised of at least 13 genera, and numerous species. Many species of this family are considered to be dangerous and several of them have caused human fatalities.

Tiger Shark, *Galeocerdo cuvieri* (Le Sueur) (Plate 2) This species is second highest in the number of shark attacks on human beings; at least 27 documented attacks are credited to it. The tiger shark is known to attain a length of at least 18 feet (5.2 meters), and probably more, but most large specimens range from 12 to 15 feet (3.6 to 4.5 meters). The distinctive characteristics of the tiger shark are its teeth, a short snout, and sharply pointed tail. Its color is gray or grayish brown, darker above than on sides and belly. The oblique or transverse blotches on the back and fins are present mostly in younger specimens, tending to fade with maturity. Large specimens are patterned only on the caudal peduncle or not at all.

The tiger shark may be sluggish, but under the proper stimulation can be a vigorous and powerful swimmer. The tiger shark is widespread throughout tropical and subtropical regions of all oceans, inshore and offshore alike, and may enter river mouths. It is said to be one of the most common of the large sharks in the tropics.

Lemon Shark, *Negaprion brevirostris* (Poey) (Figure 6). Documented human attacks have been reported for the lemon shark. It attains a length of about 11 feet (3.3 meters). Differing from other members of this family, its distinctive features are a second dorsal fin almost as large as the first, a broadly rounded snout, and its unique teeth. The lemon shark is primarily an inshore species, commonly found in saltwater creeks, bays and sounds, and around docks. This shark ranges from New Jersey to northern Brazil, and tropical West Africa, but is especially common in the waters about Florida and the Keys. It has been studied extensively because of its ability to survive in captivity.

Great Blue Shark, *Prionace glauca* (Linnaeus) (Figure 6). This shark attains a length of 12 feet (3.6 meters) or more. *Prionace* is largely a pelagic species, living in the open sea, but may also be found along coastal areas. This shark appears as a slow swimmer when not in pursuit of food, but becomes an active and aggressive fish when stimulated by food. The color of the shark is indigo blue above, slightly lighter on the sides, and white below. It is widely distributed throughout tropical and warm temperate oceans.

Bull Shark, *Carcharhinus leucas* (Valenciennes) (Figure 6). This species is believed to be synonymous with *Carcharhinus nicaraguensis* (Lake Nicaragua shark), *Carcharhinus zambezensis* (Zambezi shark), and *Carcharhinus gangeticus* (Ganges River shark), and several other closely related members—all of which are dangerous to man and have been repeatedly incriminated in human attacks. The bull shark attains a length of about 12 feet (3.6 meters) and inhabits the warmer waters of the Atlantic, Pacific and Indian Oceans. These sharks also penetrate

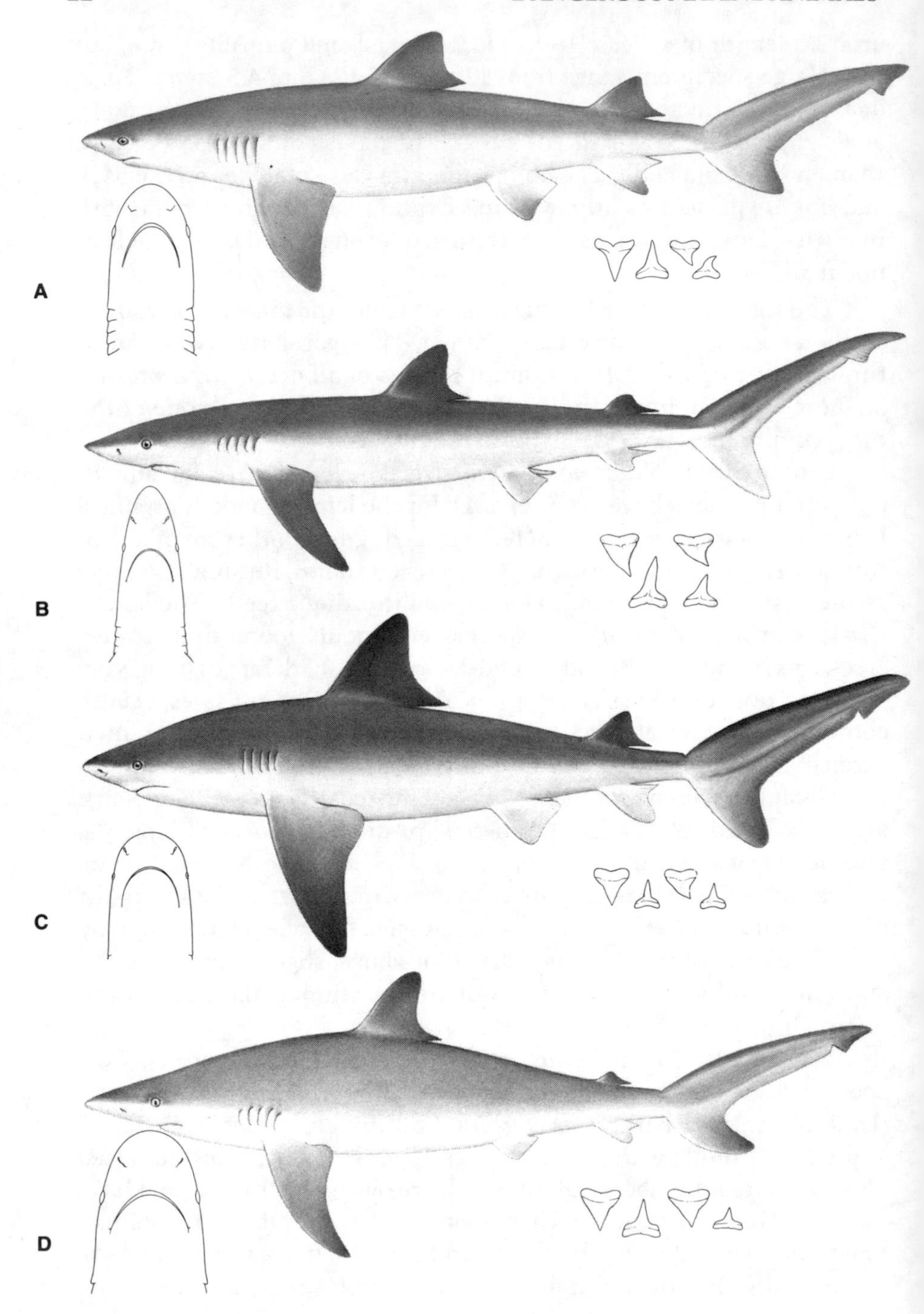
A
B
C
D

freshwater and are found in Lake Nicaraugua, Lake Izabel, Guatemala, and Lake Jamoer, New Guinea, and extend into the upper reaches of the Amazon and the rivers of Australia, Iraq, and Southeast Africa, as far as 700 miles (1120 kilometers) inland from the sea. It is usually a slow swimming shark except when stimulated by food. Gray above and white below, its fins are dark tipped in the young.

Dusky Shark, *Carcharhinus obscurus* (Le Sueur) (Figure 6). This shark attains a length of about 12 feet (3.6 meters). Although primarily a pelagic shark, it may enter shallow coastal areas. The species has been recorded from the warm Atlantic, eastern Pacific, and western Indian Oceans. It is blue gray to pale gray above and white below; the undersides and tips of the pectoral fins are grayish to dusky. The dusky shark is a seasonal migratory species.

Whitetip Oceanic Shark, *Carcharhinus longimanus* (Poey) (Figure 7). This is an active, almost fearless shark also charged in human attacks. It attains a length of about 12 feet (3.6 meters). It is widely distributed throughout the warm waters of the Atlantic, Pacific, and Indian Oceans. As a deep water species it seldom, if ever, comes close to shore. Distinctive characteristics are the broadly rounded apex of the first dorsal fin, the convexity of the posterior outline of the lower caudal lobe, and a very short snout in front of the nostrils. Colored bluish, gray, or brown above and yellowish to whitish below; its fins are white-tipped, mottled with gray. Juveniles may have dusky-tipped fins.

Bronze Whaler, *Carcharhinus ahena* (Stead). This shark, charged in human attacks, attains a length of about 9 feet (2.7 meters). An active open-ocean species, it ranges off the coast of New South Wales and southern Queensland, Australia. It is colored bronze to golden above and creamy white below. There is a closely related species, *Carcharhinus brachyurus* (Günther) (Figure 7), also known as the **Bronze Whaler**, which occurs along the coasts of New Zealand. It too is a dangerous shark.

Black Whaler, *Carcharhinus macrurus* (Ramsey and Ogilby) (Figure 7). The black whaler, charged in human attacks, attains a length of 12 feet (2.6 meters). It is found both offshore and in shallow water areas, inhabiting the waters along the east coast of Australia. It is colored

◀ Figure 6. A. **Lemon Shark,** *Negaprion brevirostris* (Poey). B. **Great Blue Shark,** *Prionace glauca* (Linnaeus) C. **Bull Shark,** *Carcharhinus leucas* (Valenciennes). D. **Dusky Shark,** *Carcharhinus obscurus* (Le Sueur).

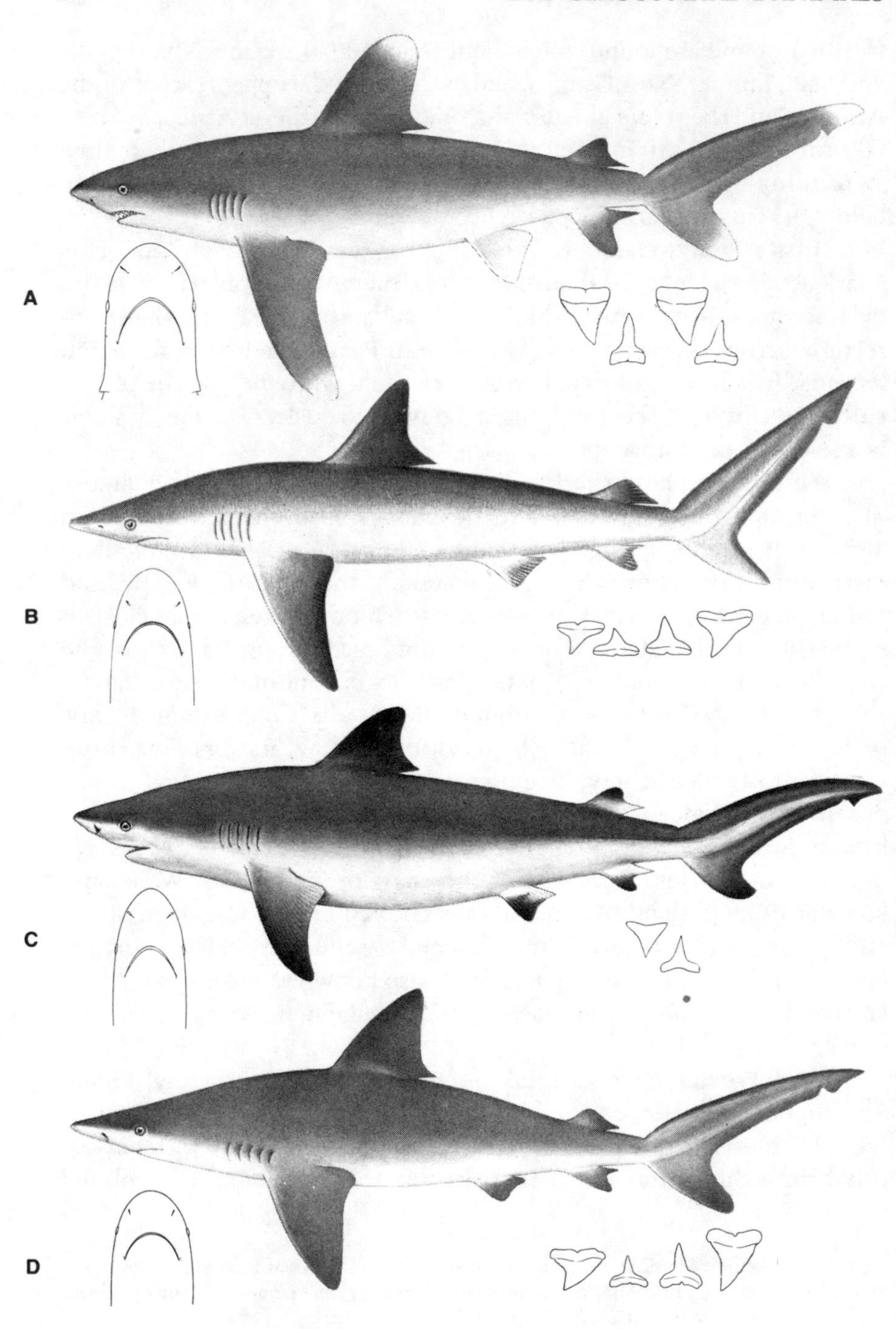
A
B
C
D

sandy and dark gray above, whitish below, and the tips of its fins are often dusky.

Whitetip Reef Shark, *Carcharhinus albimarginatus* (Rüppell) (Plate 3). Although rarely incriminated in human attacks, it is considered to be potentially dangerous. It attains a length of about 8 feet (2.4 meters). The whitetip reef shark is frequently found around reefs, coral atolls, and along shorelines. Widespread throughout the warm waters of the Pacific and Indian Oceans, and the Red Sea, this shark is distinguished by the conspicuous white markings on the first dorsal pectoral and upper caudal fins. Its colors are gray above and white below. It is sometimes confused with the whitetip oceanic shark, which differs in form and coloration, and is a deep water species.

Blacktip Reef Shark, *Carcharhinus melanopterus* (Quoy and Gaimard) (Plate 3). This shark attains a length of about 6 feet (1.8 meters) and is common about coral reefs. Widespread throughout the tropical Indo-Pacific, extending into the Red Sea, the blacktip reef shark is brownish to blackish above, paler to white below and its fin tips usually appear strikingly black. By making overt movements toward it, the shark, although charged in human attacks, can generally be frightened away.

Gray Reef Shark, *Carcharhinus amblyrhynchos* (Bleeker) (Plate 3) This species has been frequently designated as *Carcharhinus menisorrah,* which is now believed to be an entirely different species. It has been repeatedly incriminated in human attacks. Frequently found in large groups, the gray reef shark may become very aggressive if food is in the water. It can readily enter into a frenzy feeding pattern, at which time it may become quite dangerous. Attaining a length of 7 feet (2.1 meters) or more, the gray reef shark is abundant in lagoons of coral islands in the Indo-Pacific and Indian Oceans. It is grayish above, paler to whitish below, and its fins are grayish to blackish, but not black tipped.

Galapagos Shark, *Carcharhinus galapagensis* (Snodgrass and Heller) (Figure 7). Incriminated in human attacks, the Galapagos attains a length of 8 feet (2.4 meters). Found in the West Indies and in the eastern Pacific westward to the Indian Ocean, it is probably the most abundant shallow water shark in the eastern Pacific. Colored gray above and

◀ Figure 7. A. **Whitetip Oceanic Shark,** *Carcharhinus longimanus* (Poey). B. **Bronze Whaler,** *Carcharhinus brachyurus* (Günther). C. **Black Whaler,** *Carcharhinus macrurus* (Ramsey and Ogilby). D. **Galapagos Shark,** *Carcharhinus galapagensis* (Snodgras and Heller).

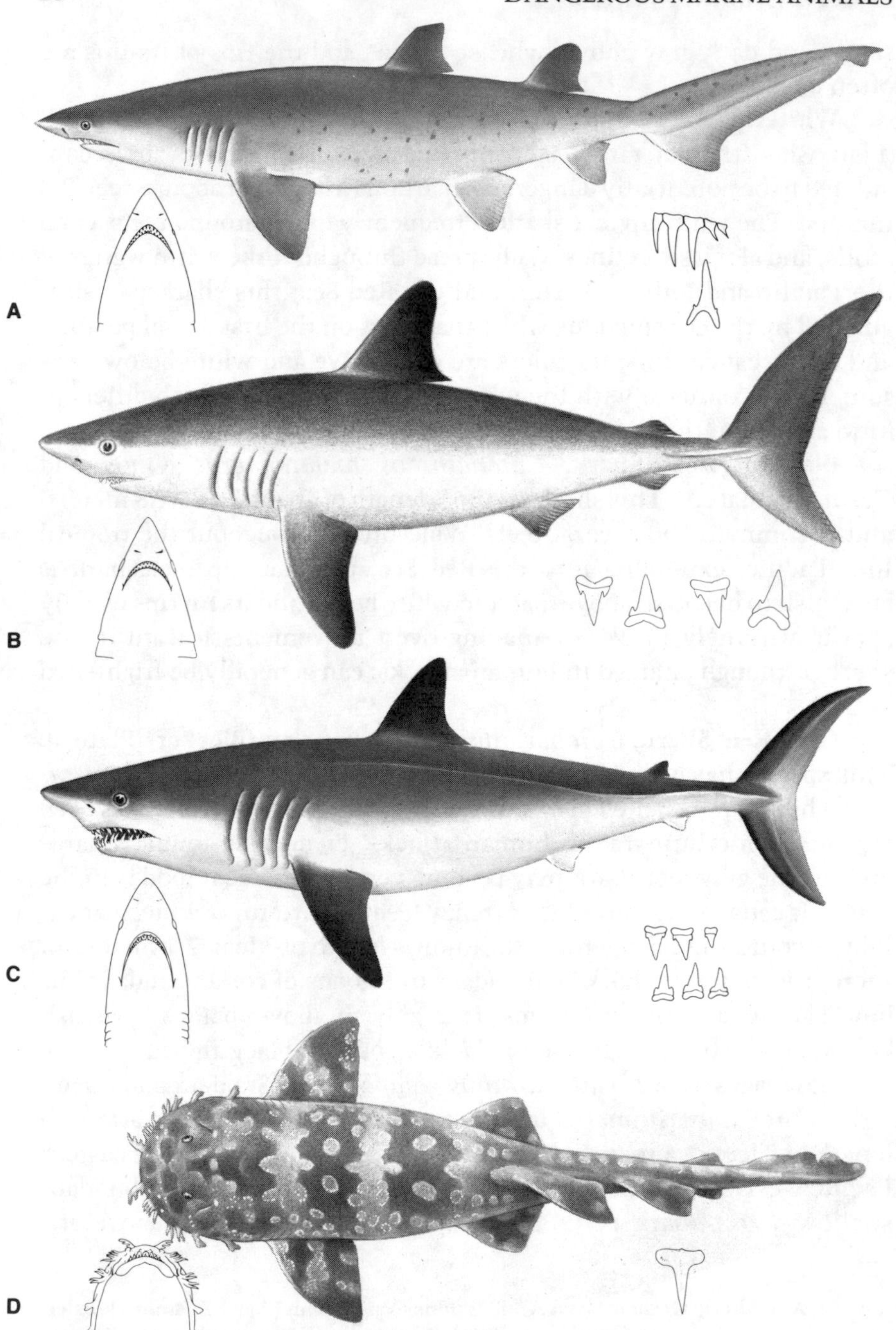
A
B
C
D

cream below, it is often mottled with gray, with the tips of the pectoral and first dorsal darker.

Family *CARCHARIIDAE*

Sand Sharks; Gray Nurse Sharks. This family includes a small group of typical sharks, having five gill openings, an anal fin, two dorsal fins of almost equal size, and no fin spines. Only one genus is known, with about six species, at least two of which have been incriminated in human attacks.

Sand Shark, *Carcharias taurus* Rafinesque (Plate 2). A comparatively sluggish shark, having a voracious appetite, it feeds very actively at night. It lives mostly near the bottom as an inshore species. The sand shark attain a length of about 10 feet (3 meters). It inhabits the Mediterranean, Atlantic, and the Indian Ocean along eastern Africa. It is gray or brown above, paler to dirty white below, and sparsely flecked on back and sides. Its fins are colored with brownish spots.

Gray Nurse Shark, *Carcharias arenarius* (Ogilby) (Figure 8). Said to be swift and savage in making its attacks, the gray nurse shark, charged in human attacks, attains a length of 15 feet (4.5 meters). As an inshore species, it lives close to the bottom of waters about Australia. Its color is similar to the *Carcharias taurus*, brownish above and dirty white below.

Family *ISURIDAE*

Mackerel Sharks; Mako Sharks. This family is comprised of only three genera, *Carcharodon*, *Lamna*, and *Isurus*, and a total of about a half dozen species or less. All the members of this family are large sharks and potentially dangerous.

Great White Shark, White Shark or **White Pointer,** *Carcharodon carcharias* (Linnaeus) (Plate 2). This shark is reputed to be the most dangerous of all sharks, despite the claim by some shark experts that the white shark is a "man biter" and not a "man-eater." It is believed by some that attacks on man are due to a case of mistaken identity. Nevertheless, this shark is credited with more attacks on humans than any other shark species. It attains a length of 20 feet (6 meters) or more. The white shark is characterized by its obtusely conical head, heavy-

◂ Figure 8. A **Gray Nurse Shark,** *Carcharias arenarius* Ogilby. B. **Mackerel Shark,** *Lamna nasus* (Bonnaterre). C. **Mako Shark,** *Isurus oxyrinchus* Rafinesque. D. **Wobbegong Shark,** *Orectolobus ornatus* (De Vis).

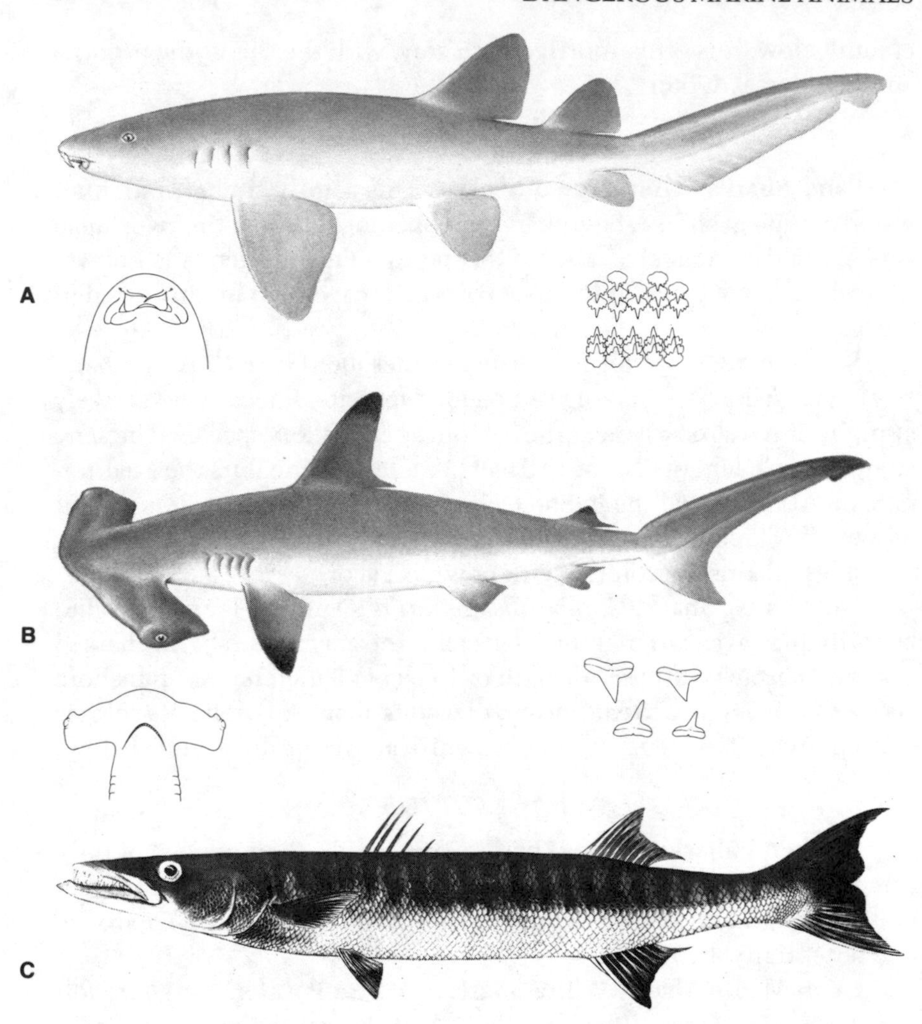

Figure 9. A. **Nurse Shark,** *Ginglymostoma cirratum* (Bonnaterre). B. **Smooth Hammerhead Shark,** *Sphyrna zygaena* (Linnaeus). C. **Great Barracuda,** *Sphyraena barracuda* (Walbaum).

bodied, strongly lunate caudal fin, and large triangular serrate teeth. Colored grayish brown, gray or bluish above, and dirty white below, it usually has a black spot on its body at the rear of the pectoral fin base. The outer tips of the pectoral fins are blackish. The great white shark ranges into the tropical and temperate oceans of the world; it is generally oceanic, but may come into shallow water. It is inclined to be solitary.

Mackerel, Porbeagle, or **Salmon Shark,** *Lamna nasus* (Bonnaterre) (Figure 8). Generally regarded as a potentialy dangerous species, it tends to be a sluggish swimmer, but may become active and aggressive when in pursuit of food. This shark attains a length of 10 feet (3 meters) or more. Inhabiting temperate regions of all oceans, it is bluish gray or gray above and white below.

Mako, Sharp-nosed Mackerel, Blue Pointer, Bonito Shark, *Isurus oxyrinchus* Rafinesque (Figure 8). A dangerous species incriminated in human attacks, this oceanic inhabitant may come close to shore. It is one of the most active of all sharks, leaping when hooked. It tends to be an aggressive and savage biter with a fearsome-looking set of jaws. It has been involved in more attacks on boats than any other shark species. Attaining a length of 12 feet (3.6 meters), it inhabits all tropical and temperate oceans. Its colors are dark blue-gray or blue above and white below.

Family *ORECTOLOBIDAE*

Carpet Sharks; Nurse Sharks; Wobbegongs. This is a large family with about 12 genera and numerous species, inhabiting the Indo-Pacific and Red Sea. They are a species of small size, inhabiting reefs and shallow water. Several species having been incriminated in attacks on humans.

Wobbegong, *Orectolobus ornatus* (DeVis) (Figure 8). This and related species have been incriminated in human attacks. Attaining a length of 7 feet (2.1 meters), the wobbegong has a prominent color pattern of blotches, spots, and marblings, together with fleshy lobes which tend to camouflage it in its shallow water habitat of weed-covered rocks. Its general color is gray or brown and it inhabits eastern and southeastern Australia.

Nurse Shark, *Ginglymostoma cirratum* (Bonnaterre) (Figure 9). This species attains a length of 14 feet (4.2 meters). It is sluggish, often found lying on the bottom in schools. Human beings come in contact with it only when inadvertently swimming near the shark or stepping on it. Nurse sharks inhabit the warm waters of the Atlantic and the Pacific Oceans. Closely related species are found in the Indian Ocean.

Family *SPHYRNIDAE*

Hammerhead Sharks. This family is characterized by their peculiar flat, wide, hammer or bonnet-shaped heads, which clearly distinguishes them from all other sharks. However, with the exception of the morphology of their head, they appear to be more closely related to carcharhinids. There are thought to be two genera, but some ichthyologists place them in a single genus, *Sphyrna,* which is comprised of about seven species. They are found in warm waters of all oceans. A number of attacks have been made on humans by hammerheads, but the precise species involved is unknown.

Smooth Hammerhead Shark, *Sphyrna zygaena* (Linnaeus) (Figure 9). The hammerhead shark tends to be a leisurely swimmer, frequently observed swimming slowly at the surface as though basking in the sun. It attains a length of about 14 feet (4.2 meters). A common species found in tropical and warm temperate waters of all oceans, its color is dark olive or brownish gray above, paler to white below with tips of the fins dusky to black.

Attack Prevention: There are numerous techniques that have been recommended over the years in an attempt to prevent shark attacks. None of the methods recommended are 100 percent effective, but it is generally conceded that there are some basic principles which, if followed, can be helpful in minimizing a nasty encounter. One must always realize that sharks are unpredictable and that every possible precautionary measure should be used when swimming in hazardous areas. It should also be kept in mind that when swimming in warm temperate or tropical oceans, sharks are likely to be present and listening, even though they may not be visible to the swimmer.

Whenever possible, swim with a group or a buddy. If isolated, you may become a prime target for an attack.

If dangerous sharks are known to be in an area, it is advisable to stay out of the water, particularly if the water is turbid. If a dangerous shark is encountered, try to exit with slow, purposeful movements rather than by a series of panicky splashes, keeping in mind that sharks are keenly sensitive to distress signals set up by low frequency vibrations in the water. If you want to attract a shark, keep splashing with erratic, irregular movements. It is advisable to face the shark at all times and not to turn your back on the beast.

Do not provoke an attack by spearing, poking, grabbing, riding, or hanging on to the tail of a shark. The skin of a shark is rough, covered with denticles similar to very coarse sandpaper, and capable of inflicting severe skin abrasions. Even a relatively small shark can inflict serious injuries when provoked. Treat large and small sharks with respect.

Blood attracts and excites sharks. Therefore, do not enter or remain in the water with a bleeding wound. Women should avoid swimming in shark infested waters during their menstrual periods.

Avoid murky or turbid water in which there is poor visibility. The swimmer has a slight edge on the situation with good visibility. Refrain from swimming far from shore in deep water, in deep water channels or in drop offs likely to be inhabited by large sharks.

Check the water carefully before jumping or diving off a boat. Avoid swimming at dusk and at night when sharks are most likely to be feeding. Do not swim in the vicinity of garbage dumps.

When spear fishing, do not tie the fish to your body. Toss the dead fish into a boat or raft, or anchor them some distance from where you are diving.

Wear dark protective clothing. Bright, shiny objects and highly contrasting clothing tend to attract sharks and barracuda.

Sharks will frequently circle their intended victim several times before coming in for the strike. Therefore, get out of the water or into a boat as rapidly as possible. Try to use a rhythmic beat with the feet, moving steadily toward the boat or shore, and try not to panic. If wearing scuba, remain submerged until reaching the boat. You are more vulnerable at the surface. When possible, move to a defensive terrain with a coral reef or some other immovable object to your back, and face the aggressor.

In the event of an air or sea disaster, do not abandon your clothing when entering the water. Clothing will not only afford some protection from abrasions, but will help to preserve body heat. Try to remain as quiet as possible in order to conserve energy. Attempt to use a life boat or a raft when possible. Do not trail arms or legs over the side of a raft. Do not fish from a raft when sharks are nearby.

Repellents: Despite the questionable value of the U.S. Navy Shark Repellent (Shark Chaser) which is comprised of a black dye and lead acetate, it should be used if available. Experimental studies have shown that some shark species will tend to avoid the dye, whereas others are unaffected by it.

If your group is threatened by a shark, form a tight circle and face the shark. If an encounter is inevitable, hit the shark on the snout or in the eye with whatever object is available. Use your hands or feet only as a last resort. A shark billy, a stick tipped with a sharp metal object such as a nail, has been found to be useful in pushing off sharks.

The value of a bang stick, smoky, or powerhead, is a matter of controversy. This is an explosive device for delivering a shot or slug to the brain or some other vital organ of the shark. Firing is accomplished by forcefully diving the triggering mechanism into the head or body of the shark. When properly applied the shot is capable of instantly killing a large shark. The problem in using a bang stick is that it should not be used by a novice. It is not only a dangerous weapon, but is only effective when dealing with a solitary shark. When working in a school of sharks it may stimulate the sharks to go into a frenzied feeding pattern. The blood and noise in the water will attract more sharks and thus compound an already precarious situation.

A CO_2 gas dart gun has recently been developed which is used primarily to disturb the buoyancy characteristics of a shark. With the use of the gas dart gun, a CO_2 charge is fired into the abdomen of the shark causing the animal to be unable to control his movements and to float helplessly to the surface. The problem with the device is that it is of questionable value in defense against a charging shark. Moreover, it is difficult to penetrate the tough hide of the shark and may aggravate the animal still further. Properly applied, it may serve as a useful defensive tool, but should be used with caution.

Electrical devices of various types have been proposed, but currently, none of them appear to be of practical value. Various poisons and drug agents are also in the experimental stage and as yet are of little value.

The use of *meshing* has been the most practical and effective method to date of controlling beach areas inhabited by dangerous sharks and visited by swimmers. This involves the use of large gill nets which are submerged parallel to the beach and close to the surf line. Sharks become immeshed in the nets and die.

Dr. Scott Johnson of the U.S. Naval Undersea Center has developed a shark screen, which is a water-filled plastic sack suspended from the surface by means of flotation rings. The sack is large enough for an adult to roll into and float loosely within the sack. The sack thus eliminates most of the teleceptor sensory stimulation of the shark and thereby provides a limited amount of protection by means of camouflage. Unfortuntely, a device such as this has limited application, but under proper circumstances would be of value.

Sonic repellent vibrations have been experimented with, but apparently are unsuccessful so far. One area that needs further examination is the possibility of developing chemical repugnatorial agents. Unfortunately, this field of research has suffered from a lack of adequate support. It is questionable that a completely effective anti-shark measure will ever be developed for the swimmer and diver.

Medical Aspects: Bites from sharks are severe and death is usually due to massive hemorrhage and shock (Figure 10). The case fatality rate on the basis of the International Shark Attack File studies involving 1,165 case histories during the years 1941 through 1968 was about 35 percent, which is considerably less than previous projections.

Treatment: The initial first aid given to the patient is of the utmost importance. Failure to apply properly these measures may result in the needless death of the victim. A person who has been bitten should be removed from the water as rapidly as possible. In some instances the control of bleeding may have to be instituted in the water if there is any unavoidable delay in removing the victim from the water. However, it should be kept in mind that blood in the water is an intense stimulation to sharks and encourages further attacks.

The patient should be treated immediately on the beach by controlling bleeding by means of a tourniquet or with the use of a pressure bandage. In applying a tourniquet the patient should be lying down and the limb elevated. A piece of material should be tied around the limb using clothing or padding to avoid pinching the skin. A stick is then passed through the knot and then twisted to tighten the tourniquet until the bleeding stops. Do not tighten the tourniquet beyond that point. Tie the stick in such a fashion that the tourniquet does not release. Loosen the tourniquet every 20 minutes. *Keep in mind that if the tourniquet is left too long the limb may become gangrenous and necessitate amputation.*

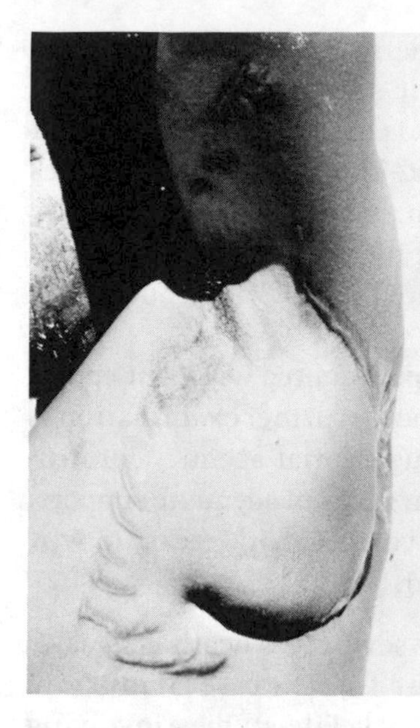

Figure 10 A. Showing scars of shark bite on a victim that survived the attack. Identification of shark unknown. Shark was believed to have been about 8 feet (2.4 meters) in length. Ninigo Islands, Bismarck Archipelago.

It is important during these procedures to continually reassure the patient. Position the patient on the sand with his head lowermost. Cover the patient lightly with a towel or thin blanket. Send for an ambulance. Do not give anything by mouth at this time. The patient should remain unmoved for about 30 minutes regardless of treatment. Over zealous rushing to the hospital can be dangerous and reduce the patient's chances of survival.

If there has been excessive bleeding, the initial intravenous therapy should be administered on the beach or boat. Intravenous solutions to be considered are Group O-Rh negative blood, plasma, 5 percent albumen, normal saline, 5 percent dextrose and water, or other plasma expanders. Attempts to control homeostasis and to combat shock are of primary concern at this time. Sedation using morphine (15 mg) or its equivalent intravenously may be required. After control of shock, the patient should be carefully transported to the hospital for further emergency treatment.

The treatment required will vary with the individual case, but there are a few general considerations of importance. Cardiopulmonary resuscitation is of primary concern—if required. Once the patient's condition has stabilized, the resulting trauma should be carefully assessed and definitive treatment instituted. Radiological examination may be required to determine the amount of bone damage. Debridement of the

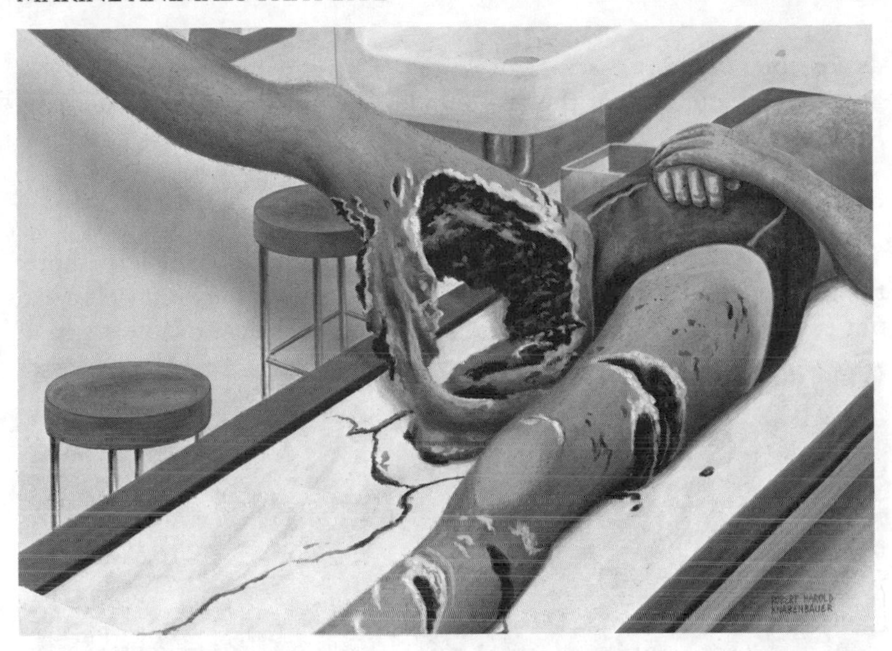

Figure 10 B. Body of 17-year-old boy killed by a **Great White Shark,** *Carcharodon carcharias.* The attack occurred on December 7, 1952, at 1400 hours, about 25 yards (23 meters) from shore, just beyond the breaker line, off the tip of Point Anlone, Monterey, California. The water temperature was 12.7°C and the water visibility was about 8 feet (2.4 meters). The boy was struck by the shark from the front with such force as to throw him out of the water to the level of his thighs. The victim was then forcefully dragged down into the water by the shark. The victim then surfaced, screaming in the midst of a bloody patch of water. The shark surfaced so as to be clearly seen by an observer on the shore and was estimated to be upwards of 13 feet (4 meters) in length. The victim was rescued by a fellow swimmer, but died on route to the shore. Examination of the body revealed that the bites inflicted were all below the hips. The major wound involved excision of the lower part of the right buttock and most of the muscle mass on the dorum of the thigh, almost to the knee. No bones were broken, but the lower part of the ischium was laid bare, and the femur was exposed for about three quarters of its length. The femoral artery was completely severed. The left leg was also severely lacerated. The boy died of massive blood and tissue loss. The drawing was prepared from the photos obtained at the time of the autopsy. This gruesome illustration graphically depicts the nature of the bite of the great white shark. This drawing and account are based on the original report by R.L. Bolin, *Pacific Science* 8:105-108, 1954.

wound is usually needed. Bacteriological cultures should be made and sensitivity tests conducted. Some species of marine bacteria may be quite virulent and resistant to most antibiotics. The wound should be photographed and documented. Conservative surgical techniques should be employed and an attempt be made to preserve as much tissue as possible. In many cases early skin grafting may be required. If available, hyperbaric oxygen therapy should be employed during the early healing period.

GIANT DEVIL RAY OR MANTA

The **Giant Ray,** *Manta birostris* (Donndorff), may attain a spread of more than 20 feet, and a weight of more than 3500 pounds (Figure 11). It does not possess a caudal sting. When harpooned, it is able to tow a small

Figure 11. **Giant Devil** or **Manta Ray,** *Manta birostris* (Donndorff). (Courtesy Dave Woodward)

launch for several hours. The manta is not aggressive and is dangerous only because of its huge size. Several deaths have been reported of helmet divers whose airlines became fouled in the cephalic fins and "wings" of mantas that apparently came over to investigate the bubbles. Like most sharks, their skin is covered by coarse dermal denticles which can inflict a severe abrasion of the skin if one were to brush up against them. They are generally seen swimming or basking near the surface of the water with the tips of their long pectoral fins curling above the surface. Occasionally, they are seen leaping out and falling back into the water with a tremendous splash. Mantas are primarily plankton feeders, but may eat crustaceans and small fishes at times. Their color is reddish or olivaceous-brown to black above, and light beneath. Mantas are found in tropical-subtropical belts in both hemispheres. Although some investigators believe that there may be as many as ten different species of mantas, they have been more recently grouped together under the single species *Manta birostris*.

BARRACUDA

Reports concerning the danger of attacks from barracuda are conflicting probably due to the fact that there are about 20 species which differ greatly in their aggressiveness. True barracuda are members of the family *Sphyraenidae* and are voracious carnivores. Some of the species are harmless because of their small size. This is true of a large school of *Sphyraena chinesis* (Lacépède), in the Marshall Islands, where divers consistently swam among them without concern, and where in no instance did they attempt to attack; nor on the other hand did the barracuda demonstrate undue fear of the divers. The same can be said for the **Pacific Barracuda,** *Sphyraena argentea* Girard.

However, under the proper circumstances, the **Great Barracuda,** *Sphyraena barracuda* (Walbaum) (See Figure 9), is said to be an exceedingly pugnacious and dangerous foe. According to E.W. Gudger (1918), who made an extensive study of the biology and habits of the great barracuda, this fish is feared more than sharks in some areas of the West Indies. Donald P. De Sylva (1963) has published an extensive review of barracuda attacks on man.

The great barracuda may attain a length of 6-8 feet (1.8-2.4 meters), and a weight of 106 pounds (48 kilograms). His mouth is large and filled with enormous knife-like canine teeth (Figure 12). A swift swimmer, striking rapidly and fiercely, *Sphyraena barracuda* is found in the West Indies and Brazil, north to Florida, and in the Indo-Pacific from the Red Sea to the Hawaiian Islands. In general, species of barracuda are widely distributed throughout tropical and subtropical waters of the world.

Skin divers report that barracuda are attracted to anything which enters the water, particularly bright-colored objects. Barracuda depend almost entirely on sight, and usually make a single attack which in most instances is non-fatal. They will follow divers by the hour, but have been seldom known to attack an underwater swimmer.

Medical Aspects: Barracuda wounds can be differentiated from those of a shark since the former are straight cuts consisting of two nearly parallel rows of tooth marks, whereas sharks produce a parabolic wound curved like the shape of their jaws (Figure 3).

Treatment: Bites should be treated in the same manner as shark bites.

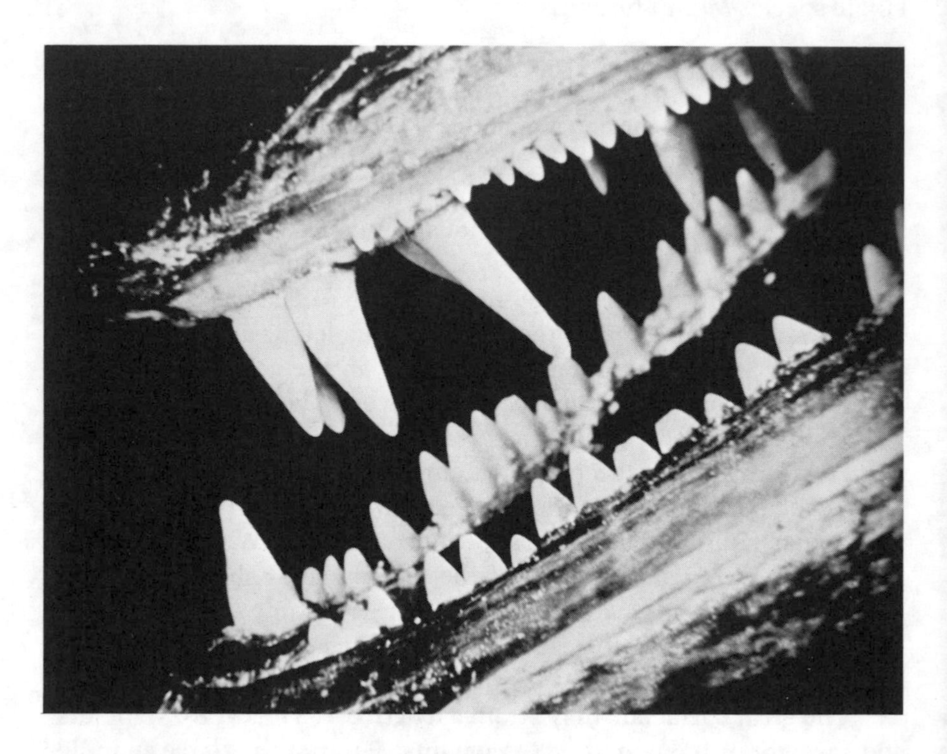

Figure 12. Jaws of the **Great Barracuda,** *Sphyaena barracuda.* Side view, from a fish about 6 feet (1.8 meters) in length. Taken at Al Ghardaqa, Red Sea.

Prevention: Barracuda are attracted by bright-colored objects and will strike at any speared fish that a diver may be carrying. When swimming in waters inhabited by the great barracuda, precautions should be taken not to attract their interest unduly. When swimming in the company of the great barracuda, treat it with respect and caution—this may be that rare one that bites.

MORAY EELS

Moray eels are members of the family *Muraenidae,* which contains some 20 or more species. They are largely confined to tropical and subtropical seas. The **Tropical Moray Eel,** *Gymnothorax javanicus* (Bleeker) is found in the Indo-Pacific (Figure 13). Several temperate zone species are known, for example *Gymnothorax mordax* (Ayres), the

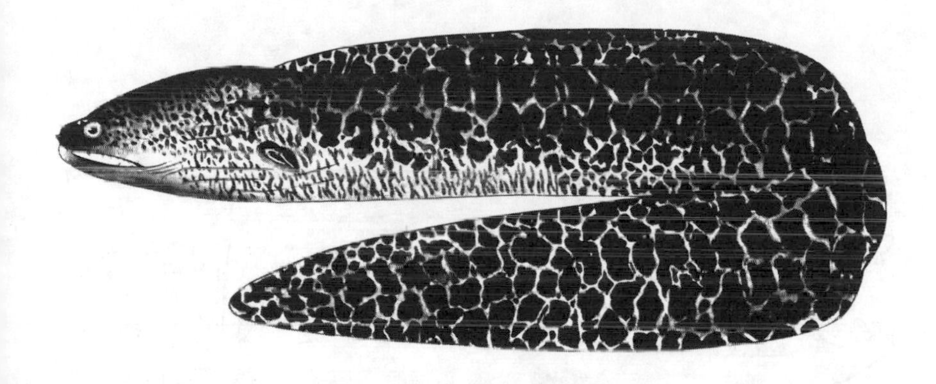

Figure 13. Tropical **Moray Eel,** *Gymnothorax javanicus* (Bleeker). Moray eels possess numerous sharp, fanglike teeth which are capable of inflicting a nasty bite. They are most likely encountered by divers reaching into holes in the reef. The flesh of some tropical species may be deadly poisonous. This species is found in the Indo Pacific region. Length 5 feet (1.5 meters).

moray eel of California, and *Muraena helena* Linnaeus of Europe. Although moray eels are notoriously powerful and vicious biters, they seldom attack unless provoked. When wounded, they can inflict severe lacerations with their narrow muscular jaws, which are armed with strong, knifelike, or crushing teeth (Plate 4 and Figure 14). They may retain their bulldoglike grip until death. Their powerful muscular bodies are covered by a tough leathery skin which is not readily penetrated by a knife. Morays are unbelievably slippery and difficult to grasp. Some of the larger species may attain a length of 10 feet, and more than a foot in diameter. Morays are bottom-dwellers, commonly found lurking in holes, and writhing snakelike through crevices, under rocks or corals.

Medical Aspects: Wounds produced by moray eels are usually of the tearing, jagged type.

Treatment: The principles involved in the treatment of shark bites are also applicable here.

Prevention: Wounds from moray eels are most likely to be encountered when poking around crevices, in holes, or under coral or rocks inhabited by them. Poke your hand into a hole with caution!

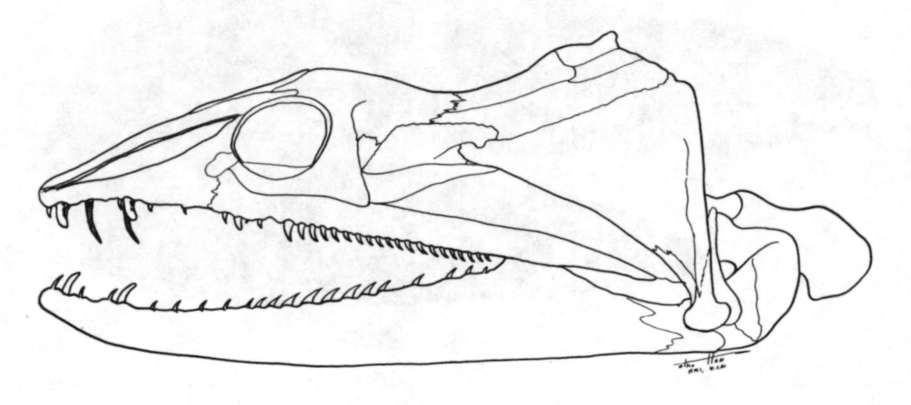

Figure 14. Skull and jaws of a typical moray eel, *Gymnothorax*, showing the sharp canine teeth. (After Gregory)

NEEDLEFISH

The needlefish, sometimes referred to as a saltwater garfish, are members of the fish family *Belonidae*. Although needlefish superficially resemble the freshwater garfish, they are phylogenetically unrelated. Needlefish are inhabitants of tropical seas, and are found in bays, inshore areas, and at times in deeper water, but are largely surface swimmers. Needlefish have a long slender body and possess two pointed elongated jaws that are filled with sharp unequal teeth (Figure 15). Some of the larger needlefish may attain a length of 2 yards (1.8 meters). Generally, when needlefish are undisturbed, they move along with an undulating motion of the body, but when excited, can scull rapidly over the surface of the water. Needlefish are frequently attracted by light at night and may leap out of the water in the direction of the light. They have also been observed leap-frogging over objects in the water. It is during these periods of excitation that they may inflict their wounds by plunging their sharp beaks into anyone in their flight path.

Medical Aspects: Needlefish have inflicted serious injuries and even fatalities to humans. These injuries have consisted of severe puncture wounds to the chest, involving heart and lungs, abdomen, arms, legs, and neck. Recently (1977), a ten-year-old Hawaiian boy was killed

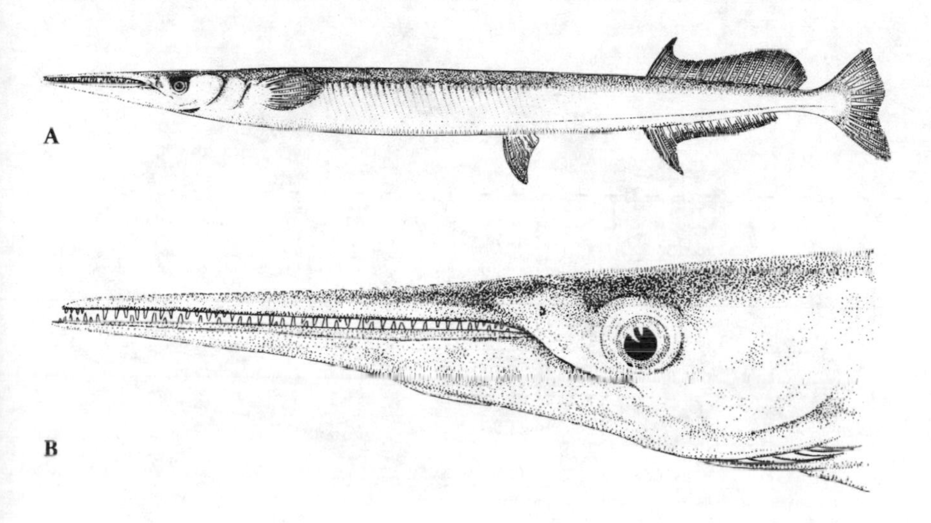

Figure 15. A. **Needlefish,** *Strongylura gigantea* (Temminck and Schlegel). B. Enlarged view of head showing the needlelike beak which is capable of inflicting a fatal wound.

while net fishing at Hanamulu Bay, Kauai, when a needlefish, believed to be *Strongylura gigantea* (Temminck and Schlegel), jumped over the net and penetrated his brain.

Treatment: Wounds are of the puncture variety and usually require prompt surgical care. The possibility of bacterial contamination must be considered. The patient should be taken to the nearest hospital or physician for treatment.

Prevention: Needlefish injuries are most likely to be encountered when night light fishing. Avoid being in a flight pattern between needlefish and the light. A person should be constantly aware of the potential danger from these fish when working at the surface in tropical waters.

GIANT GROUPER OR SEABASS

Grouper or **Seabass**, as they are sometimes called, are members of the family *Serranidae* (Figure 16). Some of the larger species may attain a length of 12 feet (3.6 meters), and a weight of more than 500 pounds (227 kilograms). There are several records in which divers have encountered difficulty from these large creatures. Grouper tend to be unusually

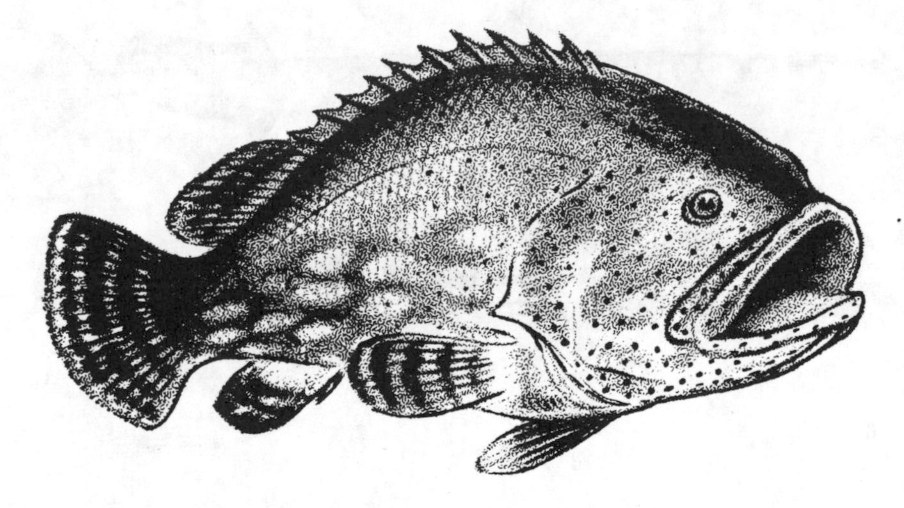

Figure 16. **West Indian Giant Grouper,** *Promicrops itaiara* (Lichenstein).

curious, bold, and voracious feeders. They are considered here as a potential hazard primarily because of their large size, cavernous jaws, and fearless attitudes. Certainly, they do not constitute a hazard in the sense that sharks do. Grouper are most frequently found lurking around rocks, caverns, old wrecks, etc.

Medical Aspects: Wounds produced by grouper would be similar to shark bites, but probably with less tissue loss.

Treatment: Similar to handling shark bites.

Prevention: Examine large caves or caverns before entering.

SALTWATER OR ESTUARINE CROCODILE

The **Saltwater** or **Estuarine Crocodile**, *Crocodylus porosus* (Schneider) (Figure 17) is one of the most dangerous of marine organisms. It ranges over an extensive geographical area including India, Sri Lanka, southern China, Western Islands, Malay Archipelago, Palau, Solomon Islands, and northern Australia. Although it usually inhabits coastal mangrove swamps, river mouths, and brackish inlets, the vast range of this crocodile is the result of its extensive excursions out to sea. Adult crocodiles have been observed swimming at sea, completely out of sight of any land.

Figure 17. **Saltwater Crocodile,** *Crocodylus porosus* (Schneider).

The saltwater crocodile may attain a length of more than 20 feet (6 meters). The back of this species is colored a brassy yellow, spotted, and blotched with black. However, older specimens may appear to be uniformly black on the back. The ventral surface is whitish or pale yellow. The stable diet of the beast is terrapins, frogs, fishes, prawns, crabs, and even water beetles. At night crocodiles enter rice paddy fields and marshy lands in search of prey, and during these excursions, it frequently captures large mammals, in some instances human beings.

The ferociousness of this crocodile is well-documented. In areas where it is endemic, it has taken numerous human lives. The crocodile frequently attacks with a loud hissing sound, grasps the victim with a powerful bite, and, in the process of dismembering the victim, drowns it with a quick twirling movement of the body. It may also sweep an animal off its feet by a blow of the tail.

Medical Aspects: Human deaths are reported each year in areas where this reptile is abundant. Once the victim is seized, only rarely is escape possible. Crocodile bites are usually serious, extensive in nature, and result in massive tissue loss. Oftentimes crocodile bites are more extensive than many shark bites. Death is usually due to the severity of the wounds inflicted and drowning.

Figure 18. **California Sea Lion,** *Zalophus californianus* (Lesson).

Treatment: Treatment is seldom possible, but if the opportunity presents itself, treat the patient in the same manner as shark bites. It should be kept in mind that crocodile wounds are frequently contaminated by bacterial pathogens, and if not treated properly may result in gas gangrene or tetanus.

Prevention: Avoid swimming in murky, brackish water inlets, swamps, and river mouths inhabited by saltwater crocodiles. Crocodiles are extremely fast on their feet and are excellent swimmers. They can inflict a lethal blow with their tails. Avoid all possible contact with these savage brutes in either daylight or dark.

SEA LIONS

Under ordinary circumstances, sea lions (Figure 18) are usually not aggressive. However, during the breeding season, the large bulls become irritable and may attack a diver, particularly if provoked. Bites from sea lions have been reported.

SEALS

Seals are generally docile, but the bulls become quite irritable during the breeding season and may inflict a serious wound. Seals need to be treated with respect (Plate 4).

KILLER WHALES

The **Killer Whale,** *Orcinus orca* (Linnaeus) (Figure 19), has long been designated as a "ruthless and ferocious killer," but more recent studies apparently fail to justify this reputation. Also, people's attitudes have changed in favor of the killer whale because they have had opportunities to observe this magnificent mammal living in oceanariums. Found in all oceans ranging from tropical to polar latitudes, the killer whale is the largest and one of the more intelligent of the living dolphins. They usually travel in pods of up to 40 individuals, are predators, fast swimmers, and feed on a wide variety of marine organisms, including invertebrates, fish, birds, seals, walrus, and even some of the larger whales. Killer whales attain a length of 30 feet (9 meters). They have a powerful set of jaws equipped with a formidable array of cone-shaped teeth (Figure 19) that are directed toward the throat and are thereby adapted for grasping and holding food. The crushing power of these jaws is immense. The killer whale is quite capable of snapping a seal or porpoise in half with a single bite. Despite their playful antics in the secluded atmosphere of an oceanarium, they must continue to be regarded with considerable respect as a potentially dangerous marine animal in the wild. Most experts are convinced that killer whales do not go in search of humans for food, but, there is always the possibility the whale might mistake a human being for other food.

A

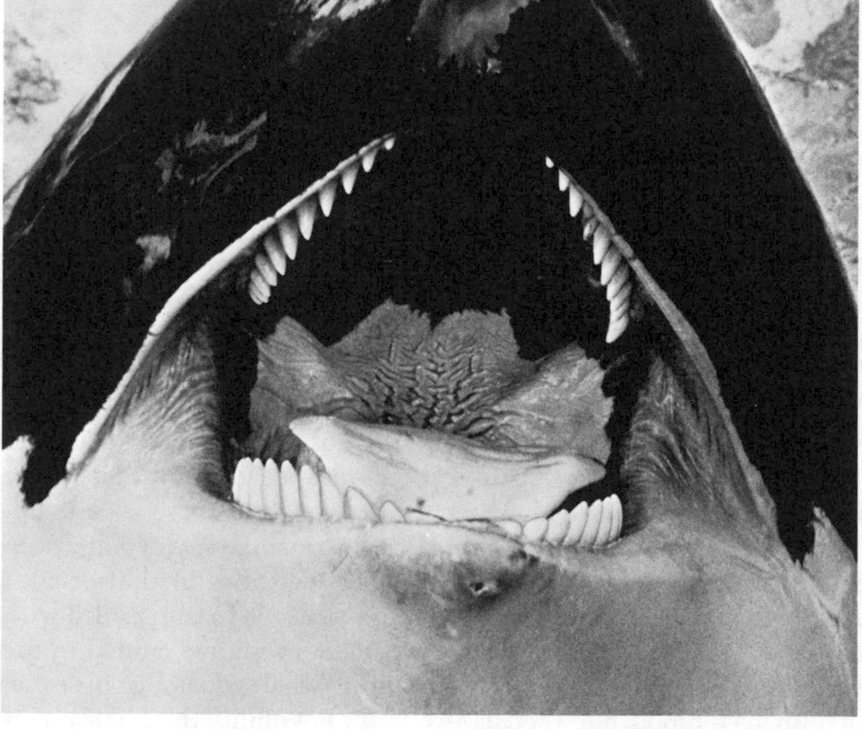

B

Figure 19. A. **Killer Whale,** *Orcinus orca* (Linnaeus). B. Looking into the mouth of the killer whale. Note the powerful set of cone-shaped teeth. The crushing power of these teeth is enormous. (Courtesy Vancouver Public Aquarium)

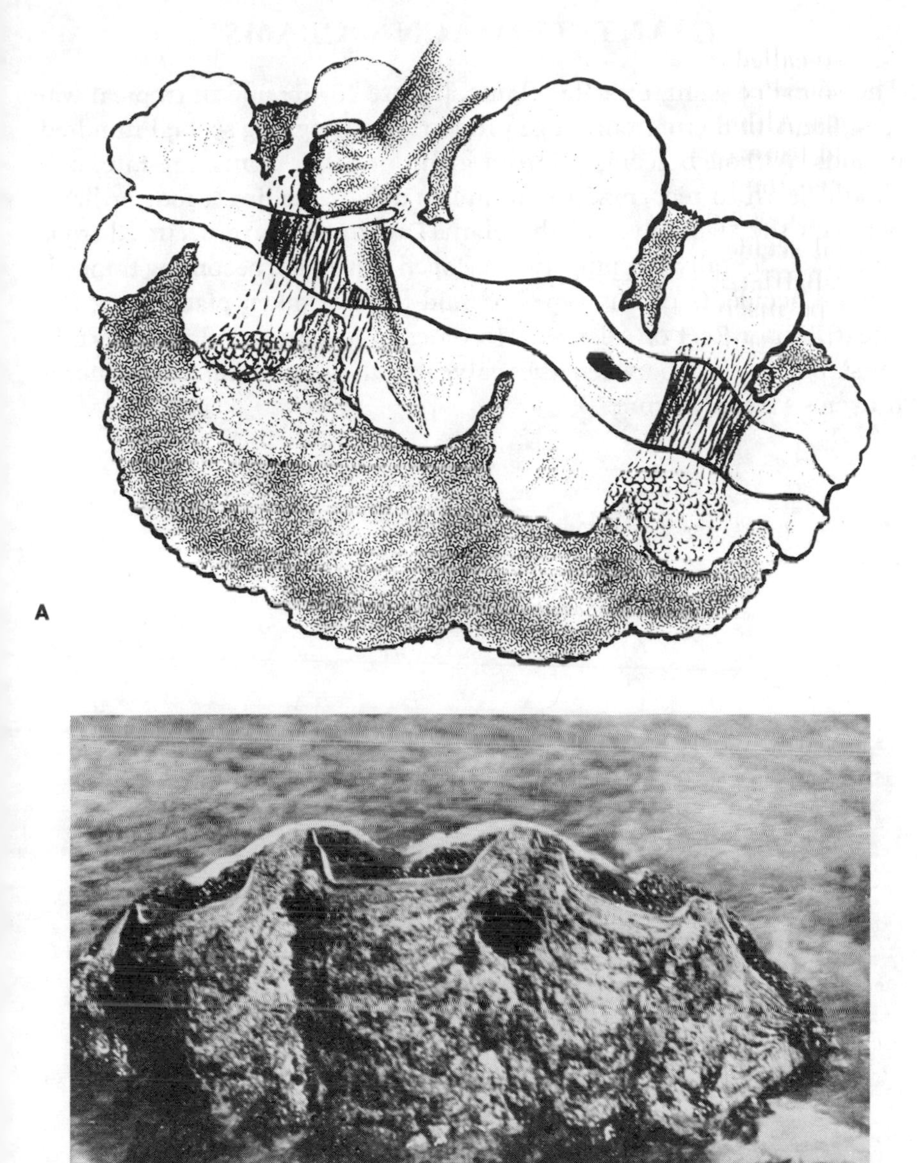

Figure 20. A. **Giant Tridacna Clam,** *Tridacna gigas* (Linnaeus). B. Photograph of *Tridacna* clam. (From Yonge)

GIANT TRIDACNA CLAMS

The so-called giant, or killer clams, (Figure 20) abound in tropical waters. Some of them attain huge proportions, weighing several hundred pounds. Although accidents from giant, or killer clams, are rare, one should learn to recognize them and to avoid catching a foot or hand between the two valves of the clam. Drownings have occurred from divers accidentally stepping into the open valves and becoming trapped. Several accidents of this type are said to have taken place along the Great Barrier Reef of Australia. In order to release the victim, a knife must be inserted between the valves of the clam and the adductor muscles severed (Figure 20).

III

MARINE ANIMALS THAT STING: INVERTEBRATES

Venomous marine invertebrates, or marine animals without backbones that inflict their injuries by stinging, can be grouped into five major categories:

1. *Porifera*—Sponges
2. *Coelenterates*—Hydroids, Jellyfishes, Corals, and Sea Anemones
3. *Molluscs*—Univalave shellfish and Octopuses
4. *Annelid Worms*—Stinging or Bristle Worms
5. *Echinoderms*—Sea Urchins

SPONGES, *PORIFERA*

Sponges are multicellular animals of simple and loose organization. Generally they have spicules of silica or calcium carbonate imbedded in their bodies for support, and fibrous skeletons made of a horny substance called spongin. It is estimated that there are about 4,000 species, some of which are found in freshwater. As a result of being exposed to chemical skin irritants present on the outer surface of the sponge, a painful dermatitis may develop from handling sponges.

Red Moss Sponge, *Microciona prolifera* (Ellis and Solander) (Plate 4). The size is variable. This sponge is found along the Atlantic coast of the United States from Cape Cod to South Carolina.

Fire Sponge, *Tedania ignis* (Duchassaing and Michelotti) (Figure 21). The size is variable and it inhabits the West Indies.

COELENTERATES, *COELENTERATA*

The coelenterates, *coelenterata*, which include the hydroids, jellyfishes, corals, and sea anemones, are simple, many-celled organisms. In

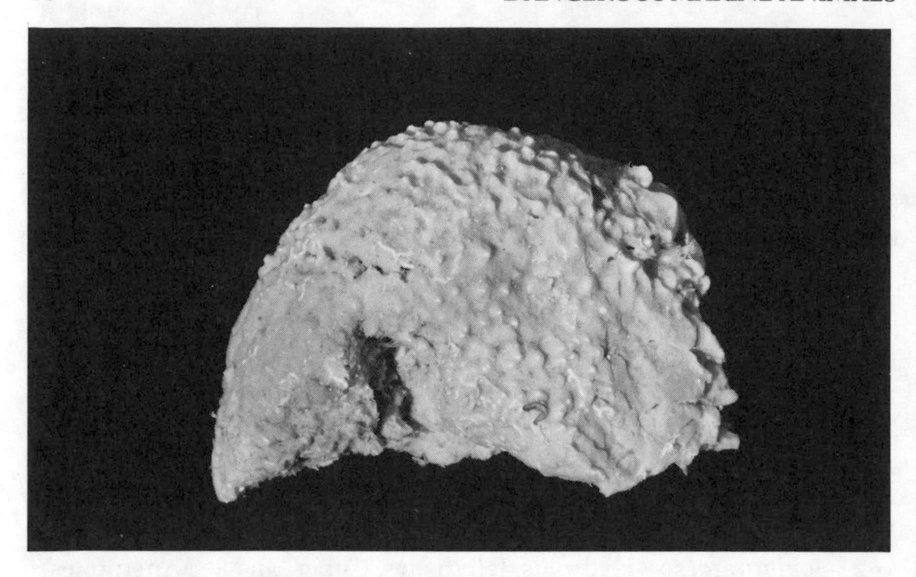

Figure 21. **Fire Sponge,** *Tedania ignis* (Duchassaing and Michelotti). The living sponges are brilliant red in color. Contact with these sponges can result in a severe dermatitis. The actual shape of these sponges may vary greatly.

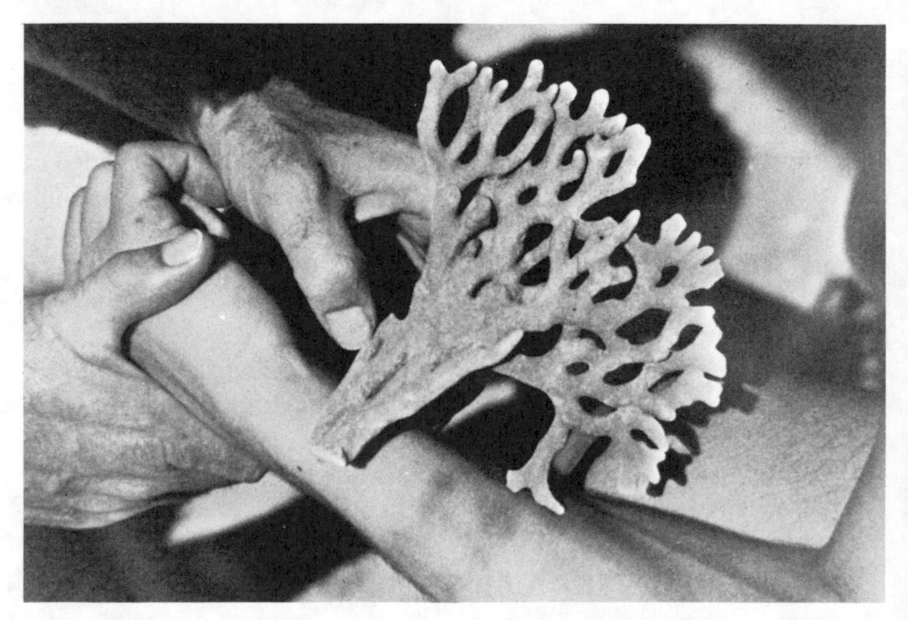

Figure 22. **Fire Coral,** *Millepora dichotoma* Forskål. Section of Fire Coral and painful rash inflicted by it. The rash subsided after about six hours. Taken at the Marine Biological Station, University of Cairo, at Al Ghardaqa, Red Sea.

Figure 23. A. **Atlantic Portuguese-Man-O-War,** *Physalia physalis* (Linnaeus). This species is found in the tropical Atlantic and throughout the West Indies. Attains a length of 13 inches (35 centimeters) or more, and tentacles may hang down into the water for a distance of more than 33 yards (30 meters). B. **Indo-Pacific Portuguese-Man-O-War,** *Physalia utriculus* (La Martinière). This smaller species is found throughout the tropical Indo-Pacific region. Attains a length of 5 inches (13 centimeters), and tentacles may hang down a distance of 13 yards (12 meters). Both species are capable of inflicting extremely painful stings.

addition to a number of other technical characters, they all possess tentacles equipped with nematocysts or stinging cells. It is the stinging cells of the coelenterates which are of concern to the skin diver.

The coelenterates are divided into three principal classes:

HYDROIDS, *HYDROZOA*

To this class belong the hydroids which are commonly found growing in plumelike tufts on rocks, seaweeds, and pilings. It is estimated that there are about 2,700 species. Some of the more common stinging members of this group are the following:

Stinging or Fire Coral, *Millepora alcicornis* (Linnaeus). This false coral is generally found living among true corals along reefs in the warm

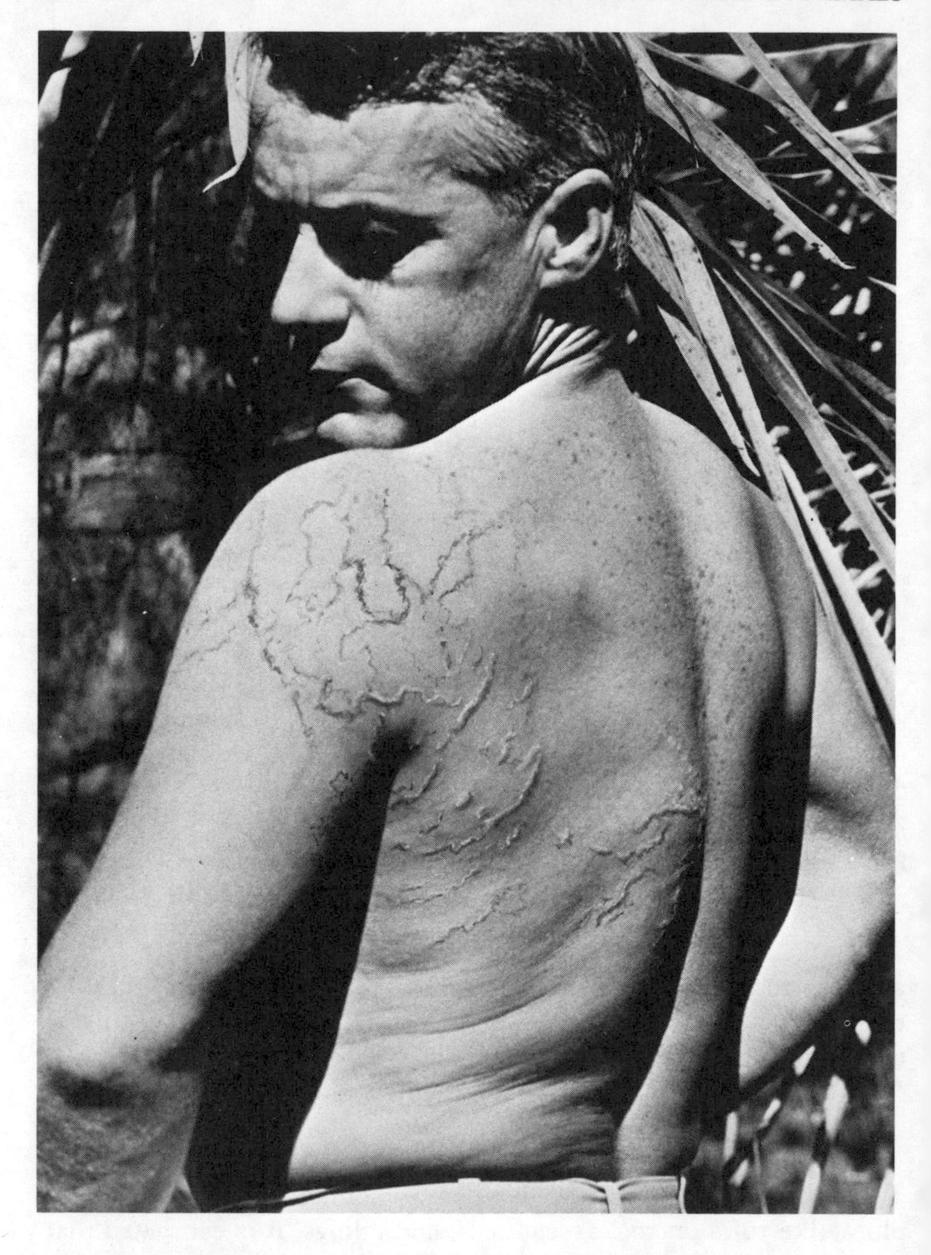

Figure 24. Sting inflicted by the **Atlantic Portuguese-Man-O-War** *Physalia physalis*. The sting caused intense pain, resulting in severe respiratory distress, and muscular paralysis, which almost caused the victim to drown. The pain lasted for about five hours and the scars for about a year. (Courtesy R. Straughan).

Plate I. **Deadly Sea Wasp**, *Chironex fleckeri* Southcott. One of the most deadly stinging creatures in the sea. Living specimen taken in an aquarium, Cairns, Queensland, Australia. Specimen about 3 inches (7.5 centimeters) across the bell.

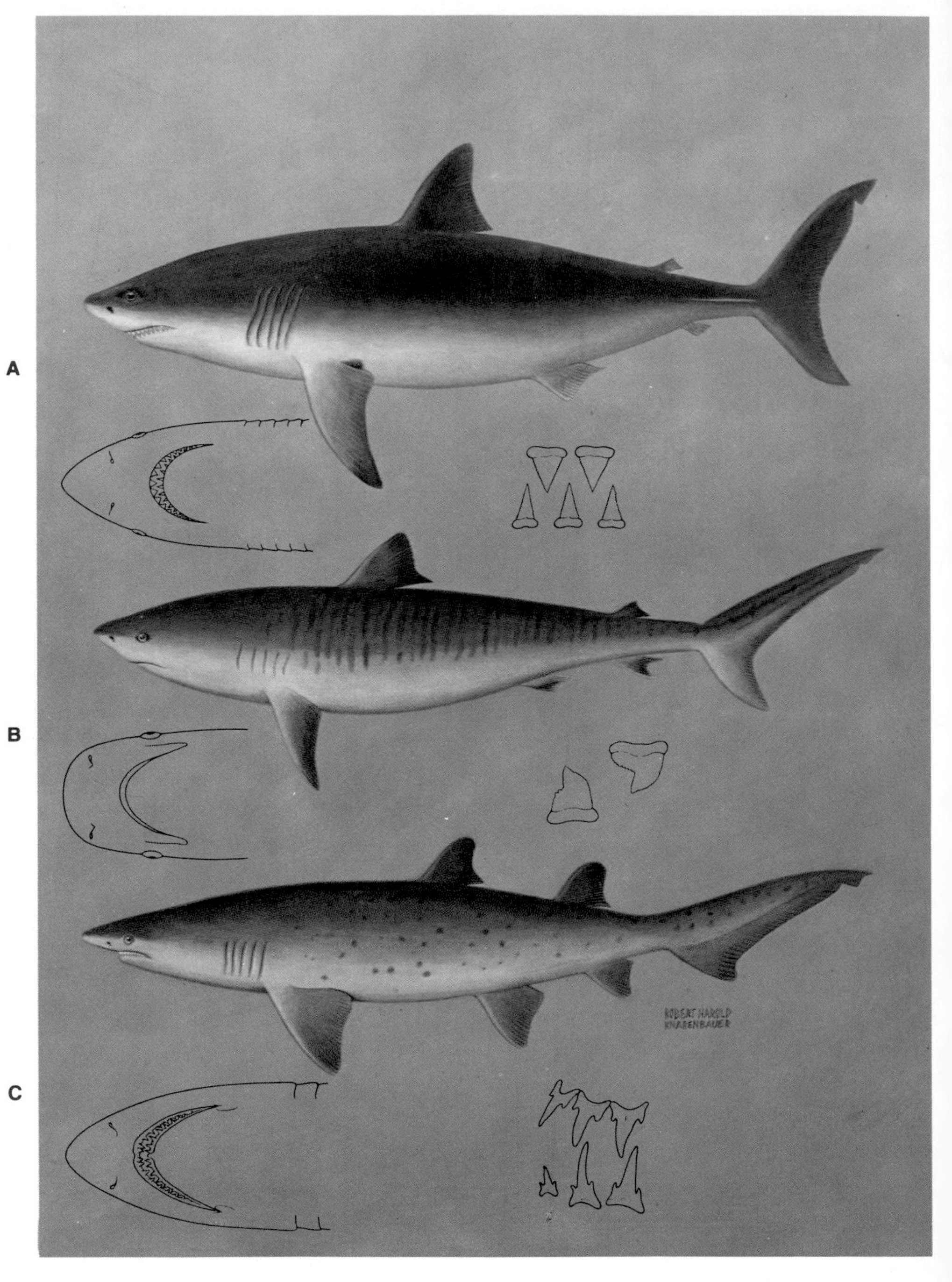

Plate 2. A. **Great White Shark**, *Carcharodon carcharias* (Linnaeus). B. **Tiger Shark**, *Galeocerdo cuvieri* (Le Sueur). C. **Sand Shark**, *Carcharias taurus* (Rafinesque). These are three of the more dangerous oceanic sharks.

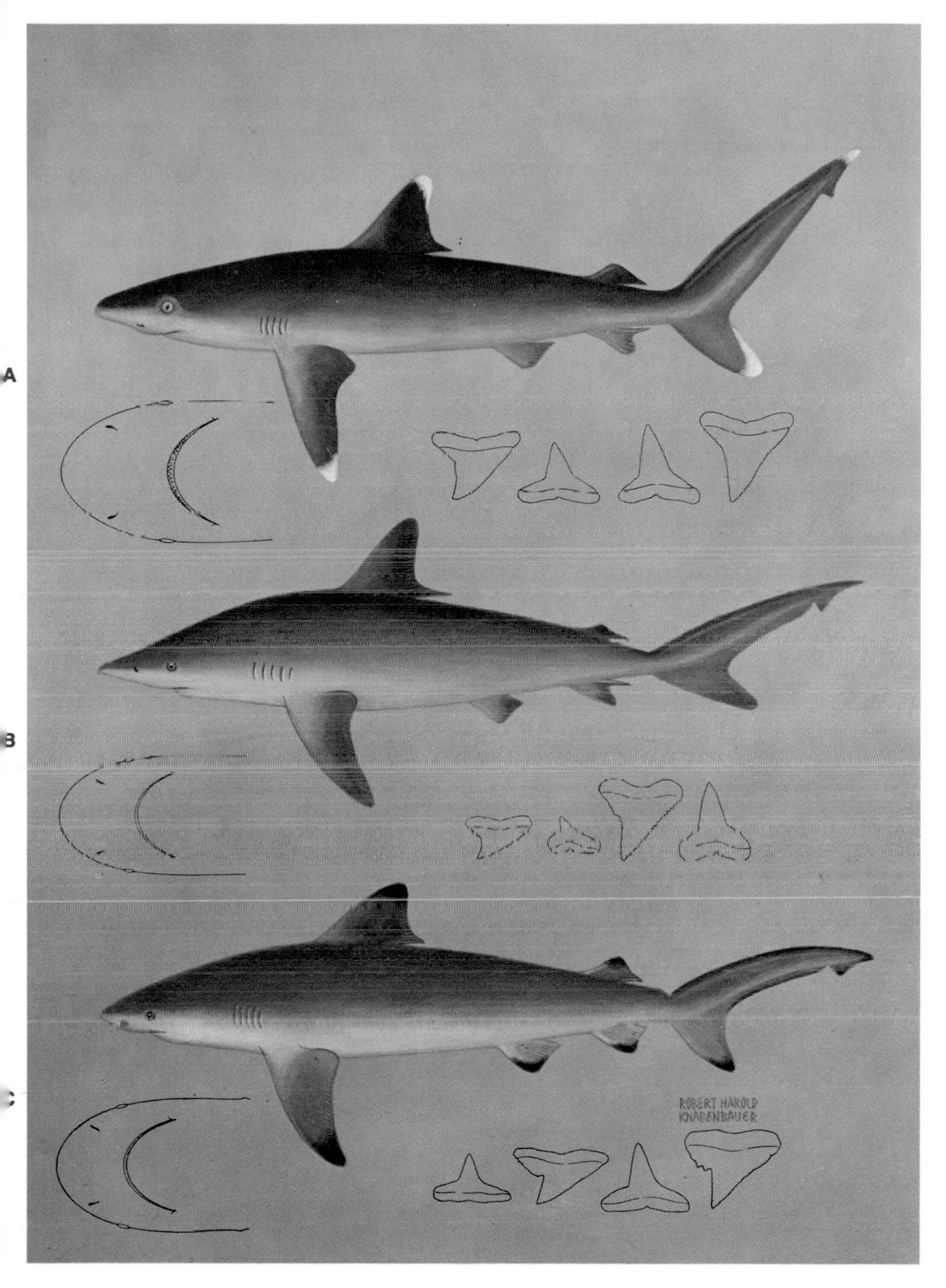

Plate 3. A. **Whitetip Reef Shark**, *Carcharhinus albimarginatus* (Rüppell). B. **Gray Reef Shark**, *Carcharhinus amblyrhynchos* (Bleeker). C. **Blacktip Reef Shark**, *Carcharhinus melanopterus* (Quoy and Gaimard). Three of the most commonly encountered, potentially dangerous reef sharks of the Indo-Pacific region. The Gray Reef Shark is the most hazardous of the three species and may be encountered either singly or in large schools.

A

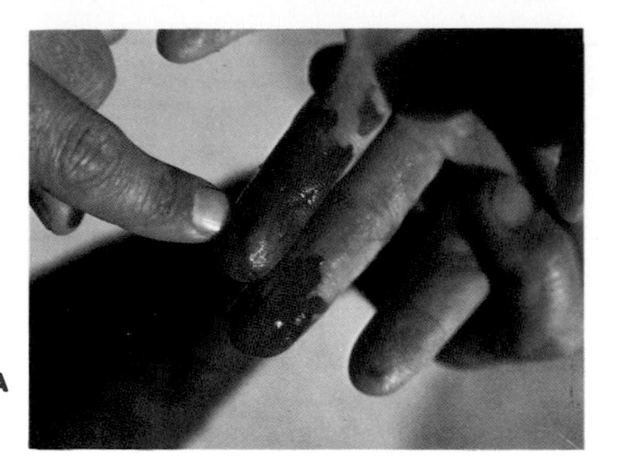

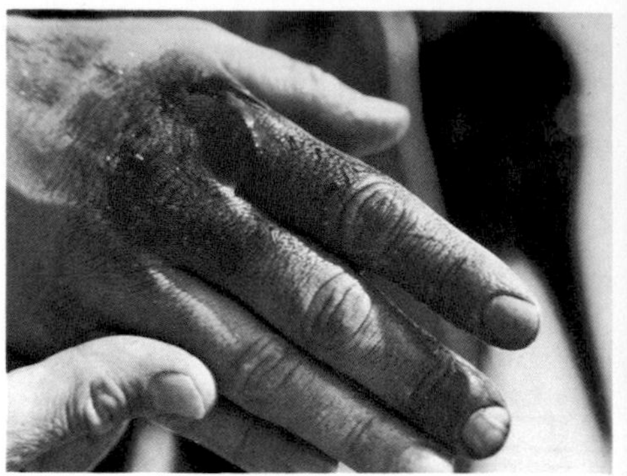

C

Plate 4. A. Seal Bite. B. Bite from a moray eel. Bites of this type are usually the result of an unsuspecting diver sticking his hand in the mouth of a moray eel. Look before you poke! This diver got off easy. Guam, Mariana Islands. C. **Red Moss Sponge**, *Microciona prolifera* (Ellis & Solander).

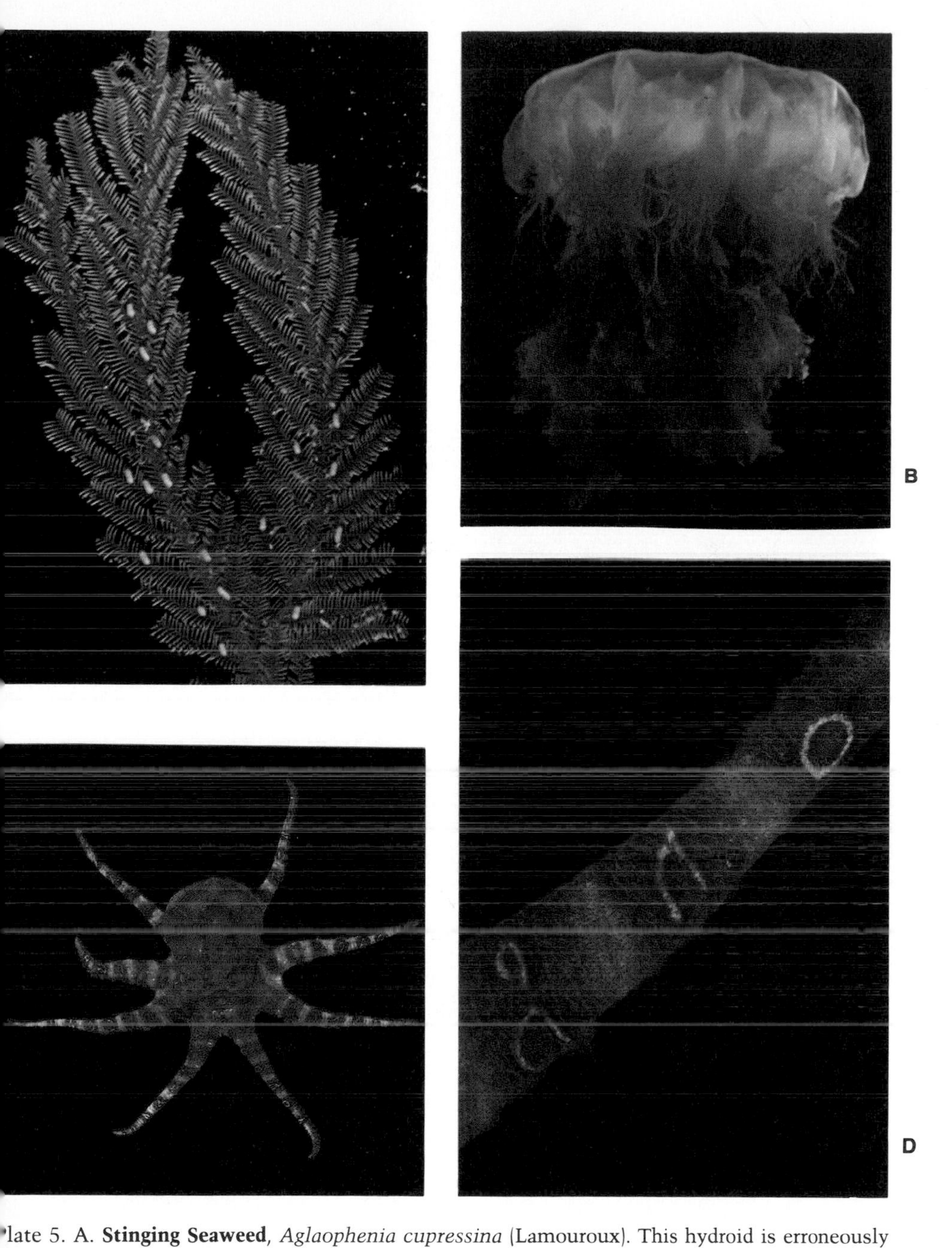

late 5. A. **Stinging Seaweed**, *Aglaophenia cupressina* (Lamouroux). This hydroid is erroneously ermed a "seaweed," but despite its general appearance, it is an animal. Drawing shows two ranches but it usually appears in large clusters. B. **Sea Blubber**, *Cyanea capillata* Eschscholtz. This s the largest of the jelly fishes and allegedly may attain a diameter across the bell of more than 5 feet .5 meters) with the tentacles dangling down more than 40 yards (36 meters). C. **Australian lue-ringed Octopus,** *Octopus maculosus* Hoyle. Living specimen taken at Sydney, New South Vales, Australia. Diameter of tentacle span, 4 inches (10 centimeters). D. Close-up tentacles of *ctopus maculosus* showing the characteristic markings. This beautiful little octopus is capable of inflicting a deadly bite.

Plate 6. A. **Sea Anemone**, *Actinia equina* Linnaeus. B. **Sea Anemone**, *Anemonia sulcata* (Pennan
C. **Sea Anemone**, *Adamsia palliata* (Bohadsch). D. **Sea Anemone**, *Sagartia elegans* (Dalyell). All these anemones are capable of inflicting painful stings.

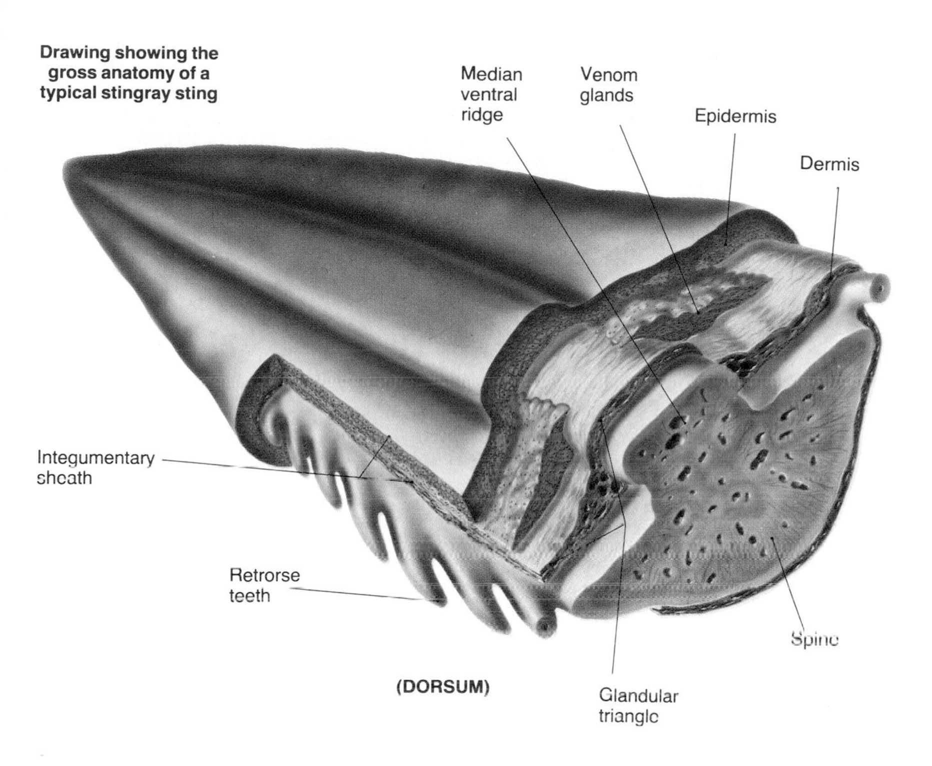

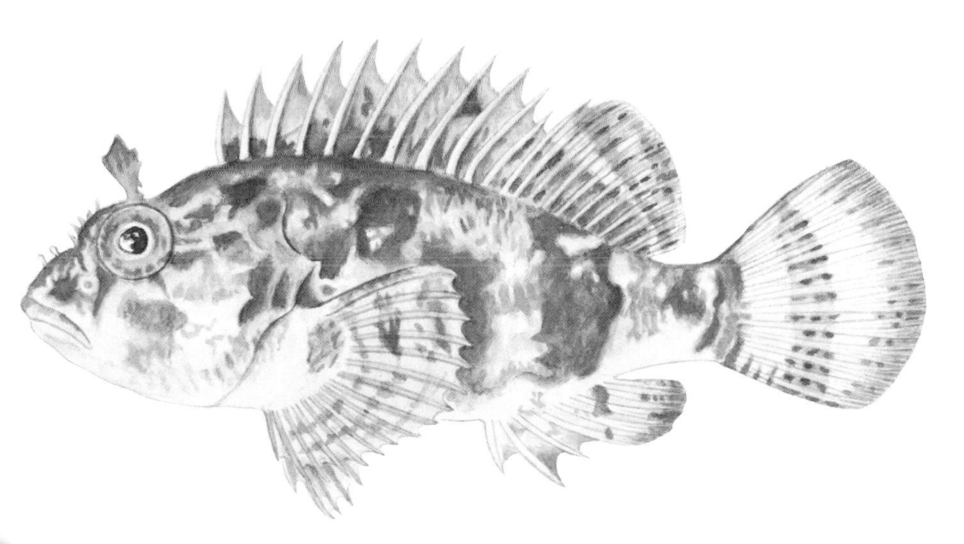

late 7. A. Drawing showing the gross anatomy of a typical stingray sting. B. **Scorpaenfish, Rascasse, Sea Pig**, *Scorpaena porcus* (Linnaeus).

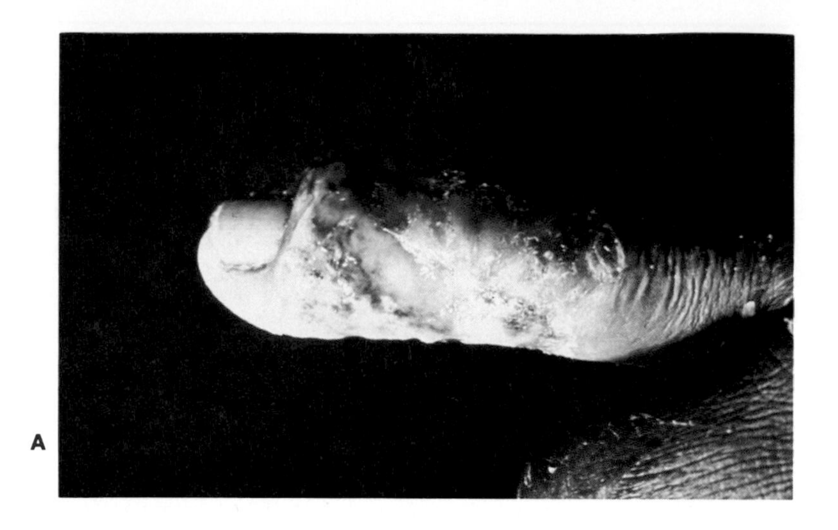

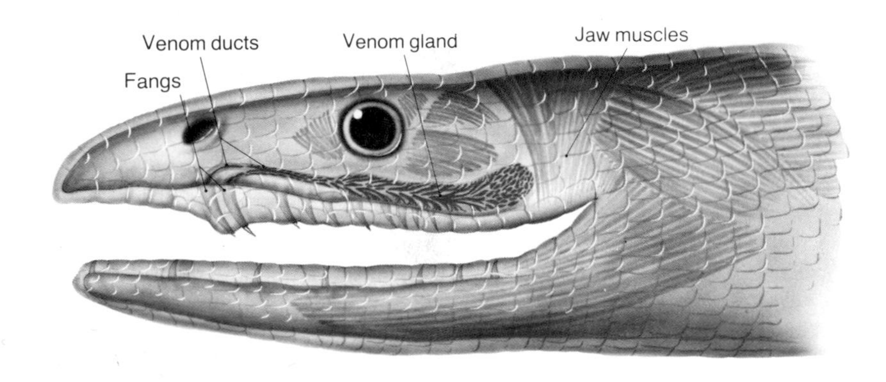

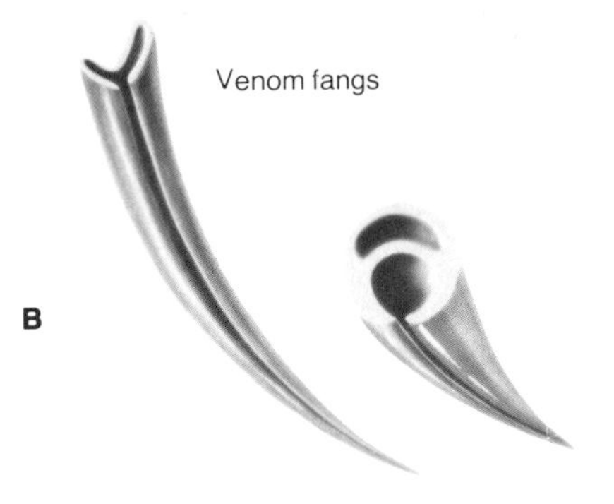

Plate 8. A. Sting inflicted by the **Stonefish**, *Synanceja verrucosa* (Bloch and Schneider), a close related species to *Synanceja horrida*. Patient had been without treatment for three weeks and t finger had become gangrenous. Guam. B. Semidiagrammatic drawing of the venom apparatus of t **Sea Snake**.

waters of the tropical Pacific, Indian Ocean, Red Sea, and Caribbean Sea. The height of the colony is variable.

Fire Coral, *Millepora dichotoma* (Forskal). (Figure 22). This species resembles *Millepora alcicornis*.

Portuguese-Man-O'-War or Blue Bottle, *Physalia physalis* (Linnaeus) (Figure 24). This hydroid is most commonly mistaken for a true jellyfish. Actually, it is a colonial hydroid. It is almost always found floating at the surface of the water. Suspended from the balloon-like floats are the stinging tentacles which may trail several feet down into the water. This particular species inhabits the tropical Atlantic, going as far north as the Bay of Fundy, the Hebrides, and the Mediterranean Sea and attains a length of 13 inches (35 centimeters) or more. A closely allied species is found in the Indo-Pacific area, Hawaii, and southern Japan and is smaller in size. Figure 24 shows the sting inflicted by the Atlantic Portuguese-Man-O'-War.

Stinging Seaweed, *Aglaophenia cupresina* (Lamouroux) (Plate 5). This hydroid has an almost seaweedlike appearance, hence its common name. It is light brown in color and capable of causing nettlelike stings with red welts and vesicles which may last several days. It is found in many parts of the tropical Indo-Pacific region. The height of the colony may reach 5 inches (12.5 centimeters).

Stinging hydroid, *Lytocarpus nuttingi* Hargitt (Figure 25). There are several closely allied species of *Lytocarpus*, all of which are capable of inflicting painful stings. Do not let their beautiful plumelike appearance deceive you—they are painful to touch. Found throughout the Indo-Pacific region, the height of the colony is about 8 inches (20 centimeters). Closely related species inhabit the West Indies.

JELLYFISHES, *SCYPHOZOA*

This class includes the larger medusae, having eight notches in the margin of the bell. It is estimated that there are about 200 species of jellyfish capable of stinging and producing mild skin irritations, but only a few might be considered especially dangerous. These include:

Deadly Sea Wasp or Box Jelly, *Chironex fleckeri* Southcott. (Plate 1). This is one of the most deadly stinging creatures in the sea. It is the cause of numerous deaths along the north Queensland coast of Australia. *Chironex* is capable of inflicting a severely lacerated rash and excruciating pain. Due to circulatory failure and respiratory paralysis,

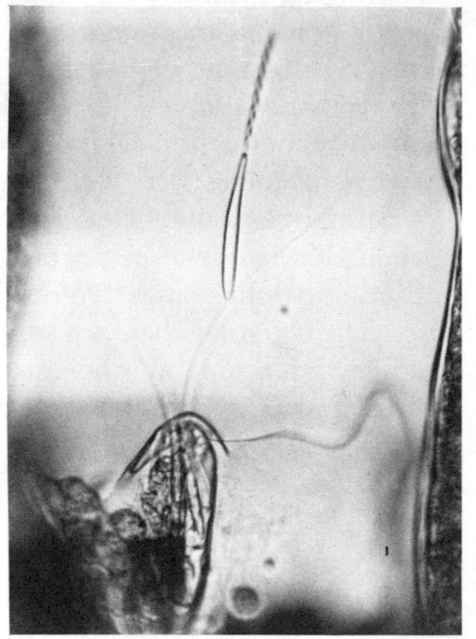

Figure 25. A. **Stinging hydroid,** *Lytocarpus nuttingi* Hargitt. Photograph taken at Ship Rock, Catalina Island, California at a depth of about 30 feet (9 meters). The featherylike organism is the hydroid, the other material is seaweed. Height of colony is up to 4 inches (10 centimeters). B. Enlarged view of one of the branches of *Lytocarpus nuttingi* which bears the polyps and stinging nematocysts. C. Photomicrograph of the nematotheca of *Lytocarpus nuttingi* with several discharged and undischarged nematocysts inside, and one discharged nematocyst outside of the nematotheca (right of center). The base of the nematocyst contains the venom. (Courtesy R. Given).

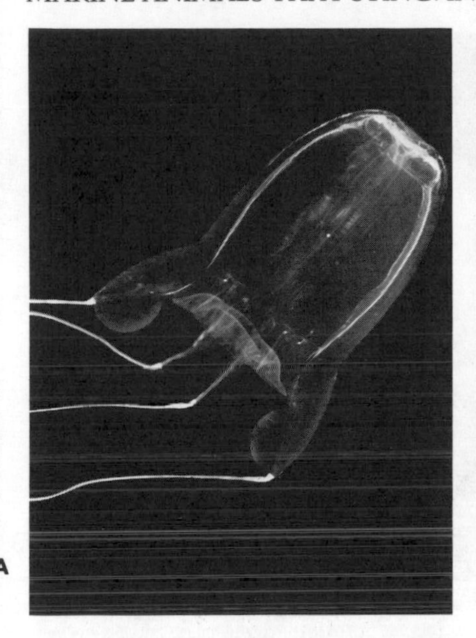

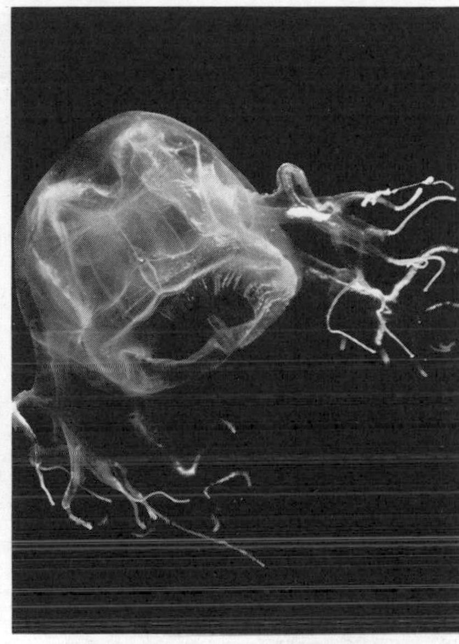

Figure 26. A. **Sea Wasp,** *Carybdea alata* Reynaud. This sea wasp is able to inflict a painful sting, but is not as dangerous as *Chironex fleckeri.* Photo of living specimen taken in an aquarium. Specimen measured about 4 inches (10 centimeters) across the bell. B. **Sea Wasp,** *Chiropsalmus quadrigatus* Haeckel. This species can also inflict a very painful sting, but is not as dangerous as *Chironex fleckeri,* with which it is sometimes confused. Specimen measured about 1.4 inches (3.5 centimeters) across the bell. Photo taken in an aquarium at Cairns, Queensland, Australia.

death may take place within a few minutes. An antivenom has now been produced for use against envenomation by this jellyfish. The antivenom is available through the Commonweath Serum Laboratories, Melbourne, Australia. A vaccine for preventive measures is in preparation. The bell may reach 4.5 inches (11 centimeters) in height.

Sea Wasp, *Carybdea alata* Reynaud (Figure 26). This species inhabits the tropical Pacific, Atlantic, and Indian Oceans.

Sea Wasp, *Chiropsalmus quadrigatus* Haeckel (Figure 26). This stinging jellyfish is an inhabitant of northern Australia, the Philippines, and the Indian Ocean. The species is sometimes confused with *Chironex,* which is much more dangerous. A closely related species occurs in the Atlantic, from Brazil to North Carolina, Indian Ocean, and northern Australia. The height of the bell is about 2 inches (5 centimeters).

A

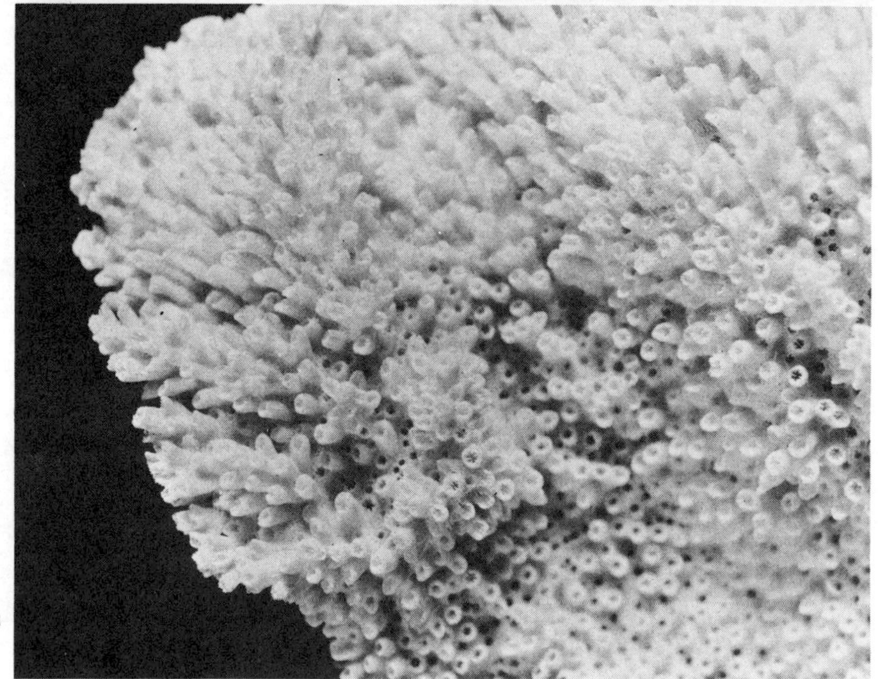

B

Sea Blubber or Lion's Mane, *Cyanea capillata* Eschscholtz (Plate 5). The sea blubber inhabits the North Atlantic and Pacific, the southern coast of new England to the Arctic Ocean, France to northern Russia, and the Baltic Sea, Alaska to Puget Sound, Japan, and China. Other stinging species of this genus are found in the tropical and temperate Pacific Ocean. They may attain large size. Some specimens have measured more than 5 feet (1.5 meters) across the bell.

SEA ANEMONES AND CORALS, *ANTHOZOA*

Elk Horn Coral, *Acropora palmata* (Lamarck) (Figure 27). Inhabits Florida Keys, Bahamas, and West Indies.

Sea Anemone, *Actinia equina* Linnaeus (Plate 6). Inhabits the eastern Atlantic from the Arctic Ocean to the Gulf of Guinea, the Mediterranean Sea, the Black Sea, and the Sea of Azov.

Rosy Anemone, *Sagartia elegans* (Dalyell) (Plate 5). Inhabits Iceland to the Atlantic coast of France, the Mediterranean Sea, and the coast of Africa.

Sea Anemone, *Adamsia palliata* (Bohadsch) (Plate 6). Ranges from Norway to Spain, and the Mediterranean Sea.

Sea Anemone, *Anemonia sulcata* (Pennant) (Plate 6). Inhabits the Eastern Atlantic, from Norway and Scotland to the Canaries, and the Mediterranean Sea.

Sea Anemone, *Triactis producta* Klunzinger (Figure 28). This is one of the more dangerous of the sea anemones, and can inflict an extremely painful sting. It inhabits the Red Sea. Its diameter is about 1.6 inches (4 centimeters), and when fully extended attains a height of about 3 inches (8 centimeters).

Hell's Fire Sea Anemone, *Actinodendron plumosum* Haddon (Figure 28). This sea anemone can inflict serious and extremely painful stings. When the sea anemone is extended, it takes on a flowerlike appearance, but when in the contracted state, it has been described as appearing to have a "top hat." Usually found on the shady side of rocks or under coral ledges, but never in the sand, the hell's fire sea anemone ranges in size from 6 inches (15 centimeters) to 12 inches (30 centime-

◂ Figure 27. A. **Elk Horn Coral,** *Acropora palmata* (Lamarck). Large branch measuring about 18 inches (45 centimeters) in height. B. Close-up of *Acropora palmata.* Corals are razor-sharp and can inflict wounds which at first may appear to be insignificant, but in the tropics are slow to heal and may result in serious infections. (Courtesy Smithsonian Institute).

A

Figure 28. A. **Sea Anemone,** *Triactis producta* Klunzinger. This Red Sea Sea Anemone, despite its small size, can inflict an extremely painful sting. Photo taken at night in the Eilat Gulf of Aqaba, Israel. B. **Hell's Fire Sea Anemone,** *Actinodendron plumosum* Haddon. Wounds inflicted by this sea anemone usually become ulcerated and may take several months to heal.

B

ters). It may produce ulceration at the envenomation site and severe systemic effects, such as, chills, fever, gastrointestinal upset, extreme thirst, and prostration.

Venom Apparatus of Coelenterates: The venom apparatus of coelenterates consists of the nematocysts or stinging cells which are located largely on the tentacles (Figures 29 and 30). These nematocysts are situated within the outer layer of the tissue of the tentacle. Each of the capsulelike nematocysts is contained inside a capsulelike device called the cnidoblast. Projecting at one point on the outer surface of the cnidoblast is the triggerlike cnidocil. Not all species possess a cnidocil. Some may be equipped with special chemoreceptors which are able to detect food and thereby activate the nematocyst. Contained within the fluid-filled capsular nematocyst is a hollow, coiled, thread tube. The opening through which the thread tube is everted is closed prior to discharge by a lidlike device called the operculum. The fluid within the capsule is the venom. Stimulation of the cnidocil or chemoreceptor appears to produce a change in the capsular wall of the nematocyst causing the operculum to spring open like a trap door, and the thread tube conveying the venom is everted. The sharp tip of the thread tube penetrates the skin of the victim and the venom is thereby injected. When a diver comes in contact with the tentacles of a coelenterate, he brushes up against thousands of these minute stinging nematocysts.

Medical Aspects: The symptoms produced by coelenterate stings vary according to the species, the site of the sting, and the person. In general, those caused by hydroids and hydroid corals (*Millepora*) are primarily local skin irritations. *Physalia* stings may be very painful (Figure 24). Sea anemones and true corals produce a similar reaction, but may be accompanied by general symptoms. Ulcerations of the skin may be quite severe in lesions produced by *Sagartia* and *Actinodendron*. Symptoms resulting from scyphozoans vary greatly. The sting of some scyphozoans is too mild to be noticeable, whereas *Chironex*, *Cyanea*, and *Chiropsalmus* are capable of inflicting very painful local and generalized symptoms. *Chironex* is one of the most venomous marine organisms known, and may produce death within a few minutes. The venomous properties of a coelenterate sting depend not only upon the chemical composition of the venom, but also upon the ability of the nematocyst to penetrate the skin of the victim.

Symptoms most commonly encountered vary from an immediate mild prickly, or stinging sensation like that of a nettle sting, to a burning, throbbing or shooting pain which may render the victim unconscious. In some cases, the pain is restricted to an area within the immediate vicinity of the contact, or it may radiate to the groin, abdomen, or armpit. The area coming in contact with the tentacles usually becomes reddened. This is followed by a severe inflammatory rash,

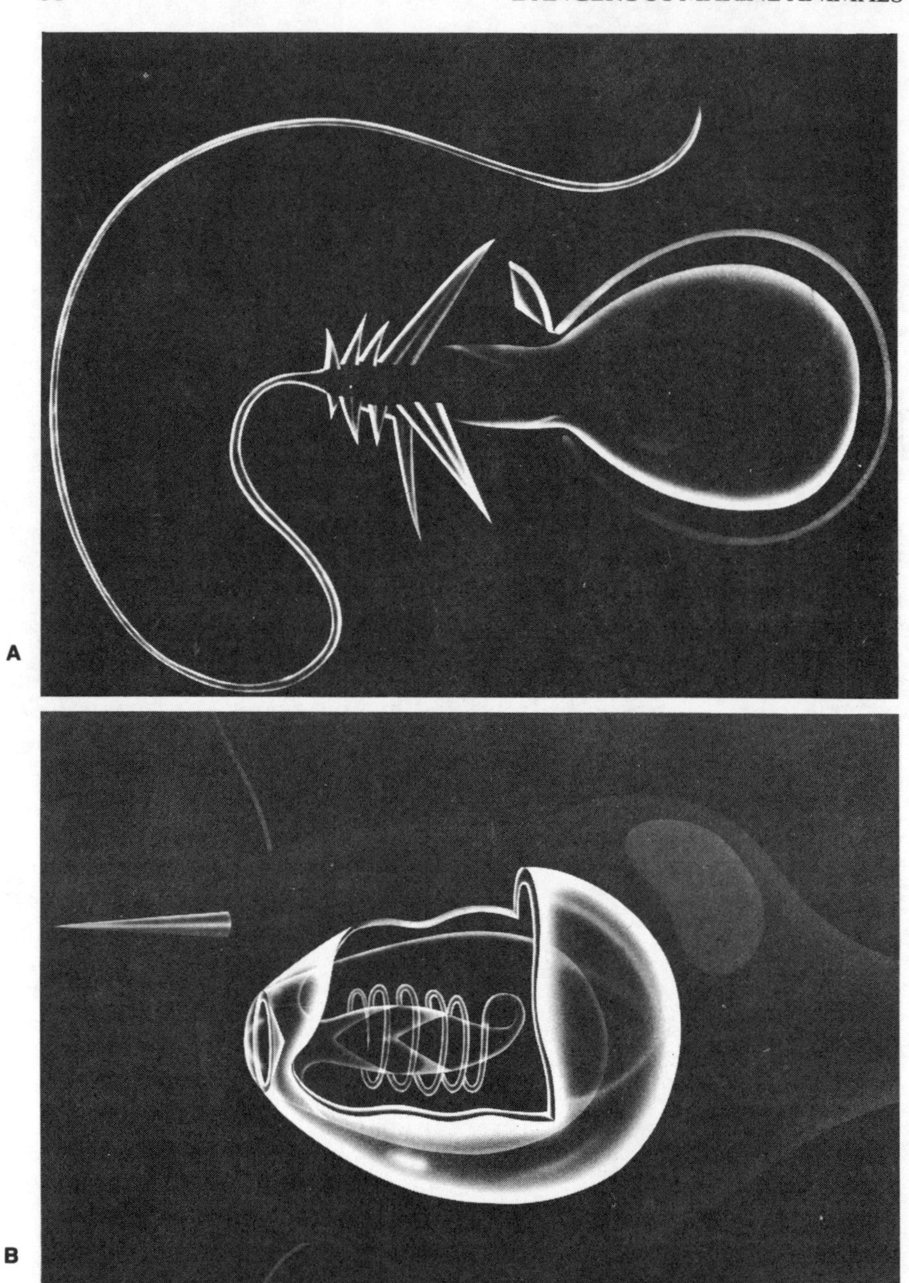

Figure 29. Nematocyst or stinging apparatus of coelenterates. A. Undischarged nematocyst. B. Discharged nematocyst. Note the coiled threadlike tube which conveys the venom. Semi-diagrammatic.

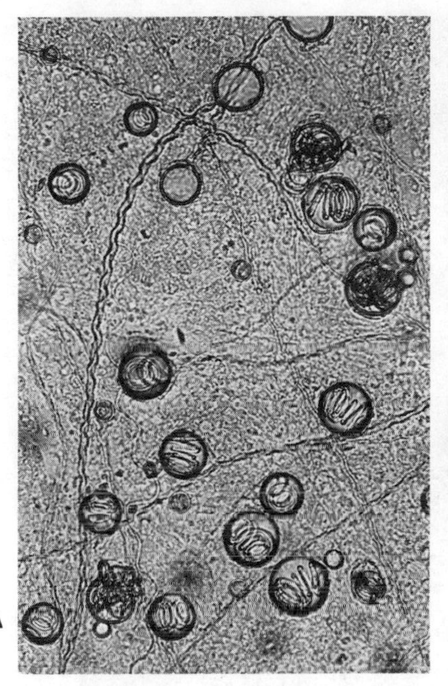

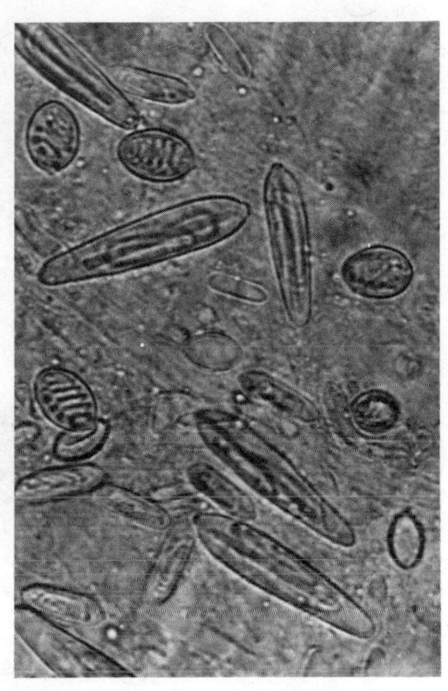

Figure 30. Photomicrographs of nematocysts: A. Nematocysts of the **Portuguese-Man-O-War,** *Physalia utriculus.* Undischarged and discharged nematocysts of both the small and large sizes are seen. (Approx. X290) (Courtesy C.E. Lane). B. Nematocysts of the **Sea Wasp,** *Chironex fleckeri.* The large undischarged nematocysts are the ones believed to be the most dangerous to man. (Approx. X240).

blistering, swelling, and minute skin hemorrhages (Figure 31). In severe cases, in addition to shock, there may be muscular cramps, abdominal rigidity, diminished touch and temperature sensation, nausea, vomiting, severe backache, loss of speech, frothing at the mouth, the sensation of constriction of the throat, respiratory difficulty, paralysis, delirium, convulsions, and death.

Treatment: Treatment must be directed toward accomplishing three objectives: relieving pain, alleviating effects of the poison, and controlling primary shock. Morphine is effective in relieving pain. Intravenous injections of calcium gluconate have been recommended for the control of muscular spasms. Oral histaminics and topical cream are useful in treating the rash. Diluted ammonium hydroxide, sodium bicarbonate, olive oil, sugar, ethyl alcohol, and other types of soothing lotions have been used with varying degrees of success. Papain (meat tenderizer) has been used with some success in the treatment of jellyfish

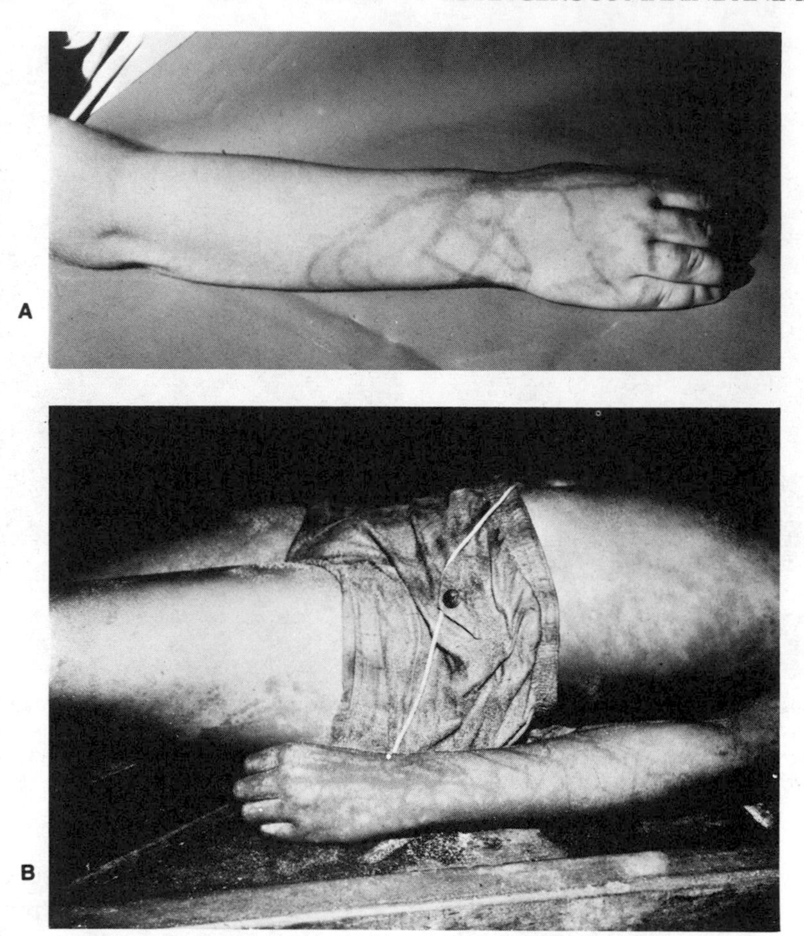

Figure 31. A. Photograph of arm of 4-year-old boy fatally stung by *Chironex fleckeri*, near Cairns, Queensland, Australia, January 31, 1966. The child was being supported by his father in waist deep water near the beach at time of stinging. Child died within 35 minutes of being stung. B. Photograph of body of 10-year-old boy fatally stung by *Chironex fleckeri*, near Mackay, Queensland, Australia, February 12, 1963. The child died within 10 minutes of being stung. The poison is a cardiotoxin, producing heart stoppage. (Courtesy J. Barnes).

stings. Prompt application of methyl or ethyl alcohol has been found to be effective in stopping envenomation by nematocysts. Keep in mind that as long as the tentacle of the jellyfish remains in contact with the skin, the nematocysts will continue to discharge their venom. It is therefore important to remove or inactivate the nematocysts as rapidly as possible. Thus the immediate application of alcohol to the skin has been found to be very effective in reducing the severity of the envenomation. Be careful about removing the tentacles with your bare fingers as further envenomation will take place. Use a stick, sand, cloth, or seaweed to remove the tentacles if nothing else is available. Artificial respiration, cardiac and respiratory stimulants, and other forms of supportive measures may be required. The only antivenom generally available for any coelenterate envenomation for treatment of the *Chironex fleckeri* is produced by the Commonwealth Serum Laboratories, Melbourne, Australia. It may be useful in treating other types of coelenterate stings because the venom of many species appear to be chemically similar in composition.

Prevention: It is important to bear in mind that the tentacles of some species of jellyfish may trail a great distance from the body of the animal—as much as 50 feet or more, in some instances. Consequently, jellyfish should be given wide berth. Tight-fitting nylon or woolen underwear or rubber skin diving suits are useful in affording protection from attacks from these creatures. Divers working in tropical waters should be completely clothed. Even though appearing dead, jellyfish washed up on the beach may be quite capable of inflicting a serious sting. The tentacles of some jellyfish may cling to the skin. Care should be exercised in the removal of the tentacles, or additional stings may result. Use a towel, rag, seaweed, stick, or handful of sand. Swimming soon after a storm in tropical waters in which large numbers of jellyfish were previously present may result in multiple severe stings from remnants of damaged tentacles floating in the water. Upon being stung, the victim should make every effort to get out of the water as quickly as possible because of the danger of drowning. Diluted ammonia and alcohol should be applied to the site of the stings as soon as possible. Rubbing the body with mineral oil, or baby oil, may help to prevent stings to some extent.

Coral Cuts: The problem of coral cuts is deserving of special mention, since they represent an ever-present annoyance to the diver working in tropical areas. Despite their delicate appearance, stony corals have calcareous outer skeletons with razor-sharp edges that are capable of inflicting nasty wounds. The severity of coral cuts is probably due to a combination of factors: mechanical laceration of the skin by the razor-sharp exoskeleton of the coral; envenomation to some extent by the nematocysts when inflicted by a living coral; introduction of foreign material into the wound (calcium carbonate, sand, debris); secondary bacterial infection; and climatic conditions (high temperature and humidity) suitable for bacterial growth. The injuries are generally superficial, but are notoriously slow in healing, and often cause temporary disability. When working in tropical reef areas for extended periods of time, "minor" coral cuts can rapidly become major medical problems, frequently resulting in severe cellulitis. Consequently, coral cuts, no matter how small, must not be ignored.

Treatment: Prompt cleansing of the wound, removal of foreign particles and dead tissue, and the application of antiseptic agents are recommended. In severe cases, it may be necessary to prescribe bedrest for the patient, with elevation of the limb, kaolin poultices, magnesium sulfate in glycerin solution dressings, and antibiotics. Antihistaminic drugs given orally, or applied locally to the wound, afford relief of pain.

Prevention: When working in the vicinity of corals, take every precaution to avoid contact with them. Do not handle corals with bare hands. Wear leather or heavy cotton gloves, and rubber-soled canvas shoes, or a completely-soled flipper.

CONE SHELLS, OCTOPUSES, ETC., *MOLLUSCA*

Stinging or venomous molluscs of concern to the skin diver fall largely into two categories:

1. Gastropods or univalve molluscs.
2. Cephalopods—octopuses and squids.

UNIVALVE SNAILS, CONE SHELLS, *GASTROPODA*

The univalves include land, freshwater, and marine snails and slugs. Members of this class are characterized by a single shell, or lack of a shell. The body is usually asymmetrical in a spirally coiled shell (the slugs are an exception to this arrangement). Typicaly, there is a distinct head, with one to two pairs of tentacles, two eyes, and a large flattened fleshy foot. Only members of the genus *Conus* are of particular concern to the skin diver.

Species of Cone Shells: Univalve molluscs of the genus *Conus*, or cone shells, as they are commonly called, because of their conelike shape, are favorites of shell collectors because of their ornate and attractive patterns. There are more than 400 species of cone shells, and all of them contain a highly developed venom apparatus. Cone shells are usually found under rocks, coral, or crawling along the sand. Several of the tropical species have caused deaths. Some of the more dangerous species are listed below:

Court Cone, *Conus aulicus* Linnaeus (Figure 32). Ranges from Polynesia westward to the Indian Ocean.

Geographic Cone, *Conus geographus* Linnaeus (Figure 32). Inhabits the Indo-Pacific, from Polynesia to east Africa.

Marbled Cone, *Conus marmoreus* Linnaeus (Figure 32). Ranges from Polynesia westward to the Indian Ocean.

Striated Cone, *Conus striatus* Linnaeus (Figure 32). Inhabits the Indo-Pacific, from Australia to east Africa.

Textile Cone, *Conus textile* Linnaeus (Figure 32). Ranges from Polynesia to the Red Sea.

Tulip Cone, *Conus tulipa* Linnaeus (Figure 32). Ranges from Polynesia to the Red Sea.

Venom Apparatus of Cone Shells: The venom apparatus consists of the venom bulb, venom duct, radular sheath, and radular teeth (Figure 33). The pharynx and proboscis, which are a part of the digestive system, also play an important role as accessory organs. The venom apparatus lies in a cavity within the animal. It is believed that preparatory to stinging, the radular teeth which are housed in the radular sheath are released into the pharynx, and thence to the proboscis, where they are grasped for thrusting into the flesh of the victim. The venom, believed to be produced in the venom duct, is probably forced under pressure by contraction of the venom bulb and duct into the radular sheath, and thereby forced into the coiled radular teeth (Figure 34).

Medical Aspects: Stings produced by *Conus* are of the puncture wound variety. Localized ischemia, cyanosis, numbness in the area about the wound, or a sharp stinging or burning sensation are usually the initial symptoms. Numbness and tingling begin at the wound site and may spread rapidly, involving the entire body, but are particularly pronounced about the lips and mouth. In severe cases, paralysis may be present. Respiratory distress is usually absent. Coma may ensure, and death is said to be the result of cardiac failure. The clinical condition of the patient may get worse during the first six hours after being bitten. If the victim survives, he usually recovers within a period of 24 hours. The localized reaction may persist for several weeks.

Treatment: Cone shell stings should be treated like a snake bite, with a ligature, incision, and removal of the venom by suction. Incision and suction are of little value if more than one hour has lapsed after the wound was inflicted. The patient should rest as much as possible and be moved to a hospital as soon as possible. Respiratory distress may develop, requiring immediate mouth to mouth resuscitation until the patient can be transferred to a mechanical respirator. External cardiac massage may be needed if pulse and heart beat are absent. If in a state of shock, place the patient on his back so that his head is lower than his feet.

When the patient arrives at the hospital, cardiac massage, defibrillation, vasopressor drugs, etc., may be required. Respiratory depressants should be avoided, and respiratory stimulants are probably useless because of the neuromuscular blocking action of the cone shell venom. Unfortunately, there are no specific antivenins available for cone shell envenomations.

Prevention: Live cone shells should be handled with care, and effort should be made to avoid coming in contact with the soft parts of the animal. Do not place cone shells in pockets. They are capable of inflicting stings through clothing.

OCTOPUSES, SQUIDS, ETC., *CEPHALOPODA*

This group includes the nautilus, squid, cuttlefish, and octopus. The head is large and contains conspicuous and well-developed eyes. The mouth, armed with horny jaws and a radula, is surrounded by eight or ten tentacles equipped with numerous suckers or hooks. Rapid move-

Figure 32. A. **Court Cone,** *Conus aulicus* Linnaeus. B. **Geographic Cone,** *Conus geographus* Linnaeus. C. **Marbled Cone,** *Conus marmoreus* Linnaeus. D. **Striated Cone,** *Conus striatus* Linnaeus. E. **Textile Cone,** *Conus textile* Linnaeus. F. **Tulip Cone,** *Conus tulipa* Linnaeus. (From Hiyama). ▶

A
B
C
D
E
F

ments can be produced by expelling water from the mantle cavity through the siphon.

Few marine animals have received greater attention from fiction writers than the octopus. Truly, it is a remarkable creature, but grossly overrated as a hazard to the diver. Both curious and cautious, the octopus can hardly be considered the demon of the seas many writers make it out to be. Large specimens may exeed 25 feet (17.5 meters) in

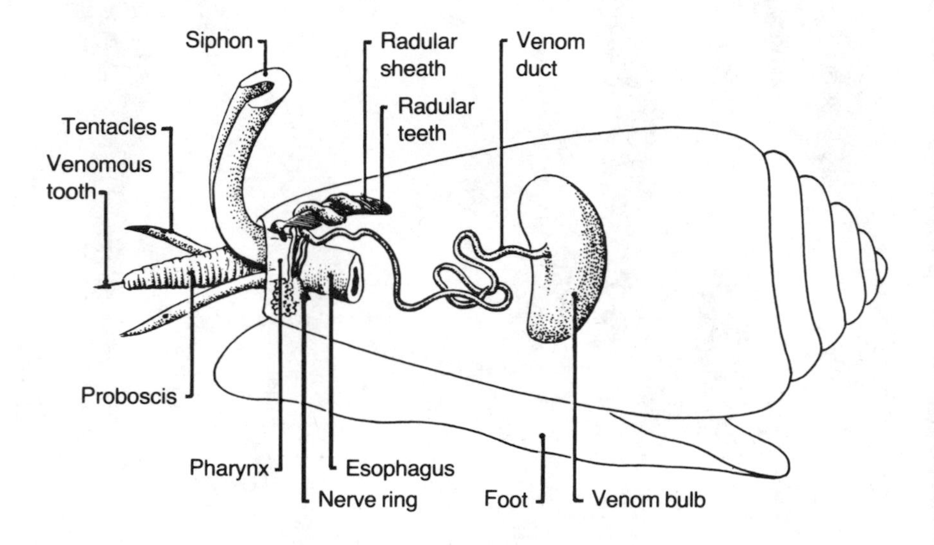

Figure 33. Venom apparatus of the Cone Shell.

span, but those encountered at diving depths are generally smaller. The giant squid that is comonly taken off the coast of South America may attain several times the length of the largest octopus, but is for the most part a deep water inhabitant, and of little concern to skin divers. The nautilus and cuttlefish are not considered to be of medical importance. Octopuses tend to hide in holes, or underwater caves. Areas such as this should be avoided by the inexperienced diver. There is a remote danger that a diver may be trapped underwater when entering a cave even by a relatively small specimen if it can get a good grip on a smooth surface of the diver. A more practical danger comes from the careless handling of small specimens of octopuses since these animals possess a well-developed venom apparatus and produce their injurious effects by biting.

Species of Cephalopods. Since cephalopods can be identified only with difficulty by specialists, no attempt will be made to consider individual species. However, one species that bears particular mention is the Australian blue-ringed octopus, *Octopus maculosus* Hoyle (Plate 5), and the closely related species *Octopus lunulatus* Quoy and Gaimard. Its length is about 8 inches (20 centimeters). Both species are venomous and capable of producing death. They are Indo-Pacific species, especially common along Southern Australian coasts, but *Octopus lunulatus* appears to have a more northerly distribution.

Venom Apparatus of the Octopus. The venom apparatus of the octopus is comprised of the so-called anterior and posterior salivary glands, the salivary ducts, the buccal mass, and the mandibles, or beak (Figure 35). The mouth of the octopus is situated in the center of the oral and anterior surface of the arms, surrounded by a circular lip, fringed with fingerlike papillae (Figure 36). The mouth leads into a pharyngeal cavity having thick muscular walls. This entire muscular complex is known as the buccal mass, which is surrounded and concealed by the muscular bases of the arms. The buccal mass is furnished with two powerful dorsal and ventral chitinous jaws, whose shape resembles that of a parrot's beak. The arrangement of the jaws differs from the parrot in that the ventral one bites outside the dorsal and is wider and larger. These jaws are able to bite vertically with great force, tearing the captured food which is held by the suckers before it is passed on to the rasping action of the radula. The duct from the posterior salivary glands opens on the tip of the sub-radular organ, which appears as an outgrowth in front of the tongue. The paired ducts from the anterior salivary glands open into the pharynx laterally and posteriorly. The venom is discharged from these ducts into the pharynx.

Medical Aspects: Octopus bites usually consist of two small puncture wounds which are produced by the sharp parrotlike chitinous jaws. A burning or tingling sensation is the usual initial symptom. At first, the discomfort is localized, but may later radiate to include the entire appendage. Bleeding is frequently profuse for the size of the bite, which may indicate that the clotting process of the blood is retarded. Swelling, redness, and heat commonly develop in the area about the wound. Recovery is generally uneventful. However, bites from the Australian blue-ringed octopus can be serious. Initially, the bite may be painless. Within a few minutes the area about the bite appears blanched and soon becomes swollen and hemorrhagic. A stinging sensation may develop shortly thereafter. The venom (maculotoxin) is a potent neurotoxin

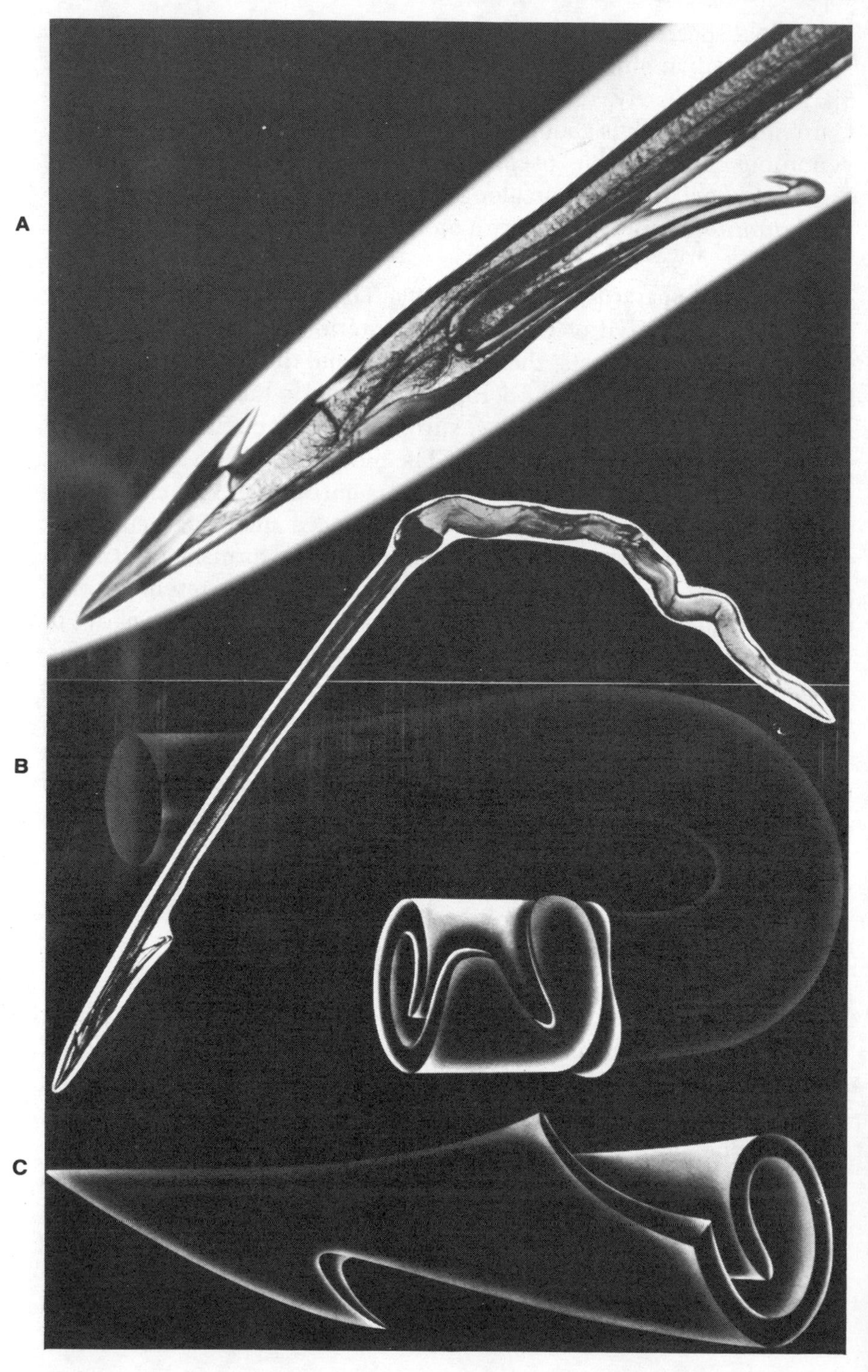
A
B
C

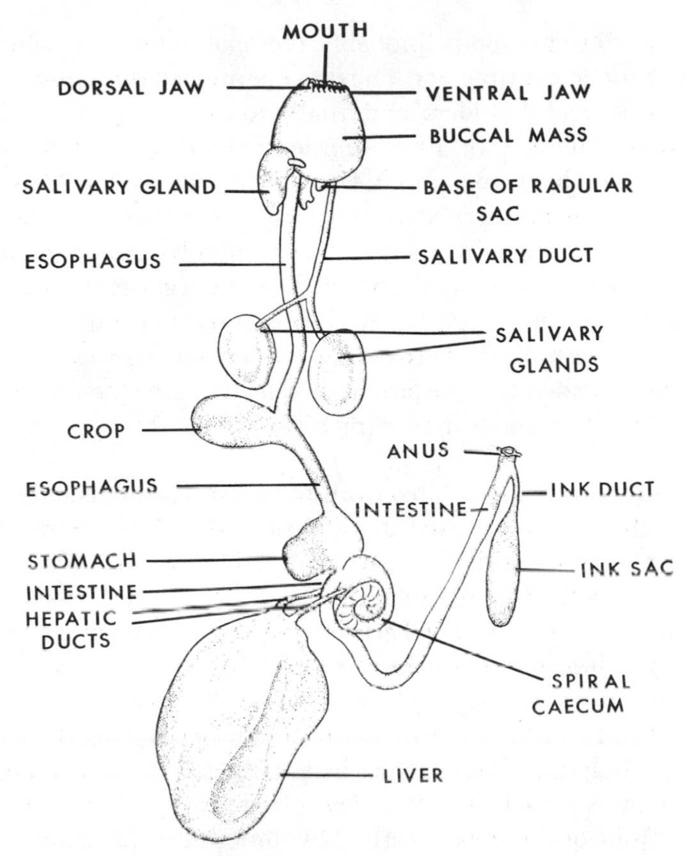

Figure 35. Anatomy of the venom apparatus of the octopus.

causing a sensation of numbness and tingling around the mouth, neck and head. This may be followed by nausea and vomiting. In severe envenomations respiratory distress soon develops, followed by visual disturbances, occular paralysis, dizziness, difficulty in speech and swallowing, lack of motor coordination, muscular weakness and, possibly, complete paraysis. In fatal cases, the patient loses consciousness, and death is due to a respiratory paralysis. If the patient survives, the duration of the paralysis is about 4 to 12 hours.

Several fatalities have been reported from octopus bites. The following report is one such example. The incident took place near East Point, Darwin, Australia. According to the account, a diver captured a

◂ Figure 34. A. Photomicrograph of the harpoonlike tip of the radular tooth of *Conus striatus*. B. Radular tooth of *Conus striatus*. C. Semidiagrammatic drawing of a typical radular tooth of *Conus*. The radular teeth are comprised of a flat sheet of chitin rolled into a tupelike structure. The venom is contained within the lumen of the tooth. Upper picture illustrates base of tooth attached to ligament. Barbed tip of hollow tooth in lower drawing.

small blue-ringed octopus (probably *Octopus lunulatus*) which had a span of about 20 centimeters. The diver permitted the octopus to crawl over his arms and shoulders, and finally to the back of his neck, where the animal remained for a few moments. During the period that the octopus was on his neck, a small bite was inflicted, producing a trickle of blood. A few minutes after the bite, the victim complained of a sensation of dryness in his mouth and of difficulty in swallowing. After walking a short distance up the beach from the scene of the accident, the victim began to vomit, developed a loss of muscular control, and finally suffered from respiratory distress, and was unable to speak. Even though he had been rushed to a hospital and placed in a respirator, the victim expired about two hours after being bitten.

Treatment: Octopus bites from most species are usually of minor concern and can be treated symptomatically. Bites from the blue-spotted octopus must be treated vigorously. Ligature, when possible, and removal of the venom by suction, or washing out the wound should be performed promptly. The patient should be placed at rest and given reassurance. If respiratory paralysis develops, mouth to mouth resuscitation and cardiac massage may be required. Attention must be given to clearing the airway of the patient from vomitus (laying the patient on his side is helpful). The head of the patient should be extended backwards as far as it will go and the jaw lifted forward, bringing the lower teeth in front of the upper teeth. This procedure will help to pull the tongue away from the back of the throat. The victim should be transported to the hospital as soon as possible. Central respiratory stimulants may be of value—especially during the recovery phase. The administration of artificial respiration is of the utmost importance. There is no specific antivenin available.

Prevention: Underwater caves which are likely to be inhabited by octopuses should be avoided by the inexperienced diver. The wearing of an outer cloth garment makes it difficult for an octopus to adhere to the skin. Regardless of their size, octopuses should be handled with gloves. Some of the smaller species seem to be the most aggressive biters. If not familiar with the species, it is best to leave it alone. Stabbing deep between the eyes is the method of choice for killing an octopus.

SEGMENTED WORMS, *ANNELIDA*

Segmented worms, or annelids, are organisms having an elongate body, which is usually segmented, each segment having paired setae, or bris-

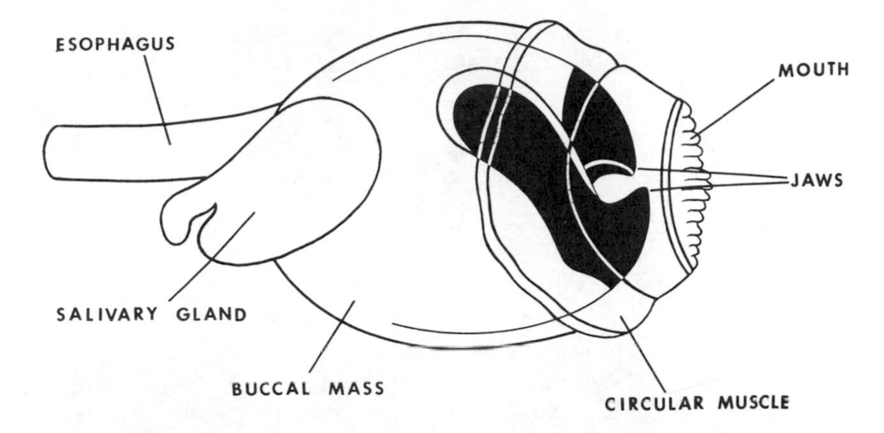

Figure 36. Buccal mass of an octopus showing the mouth and jaws.

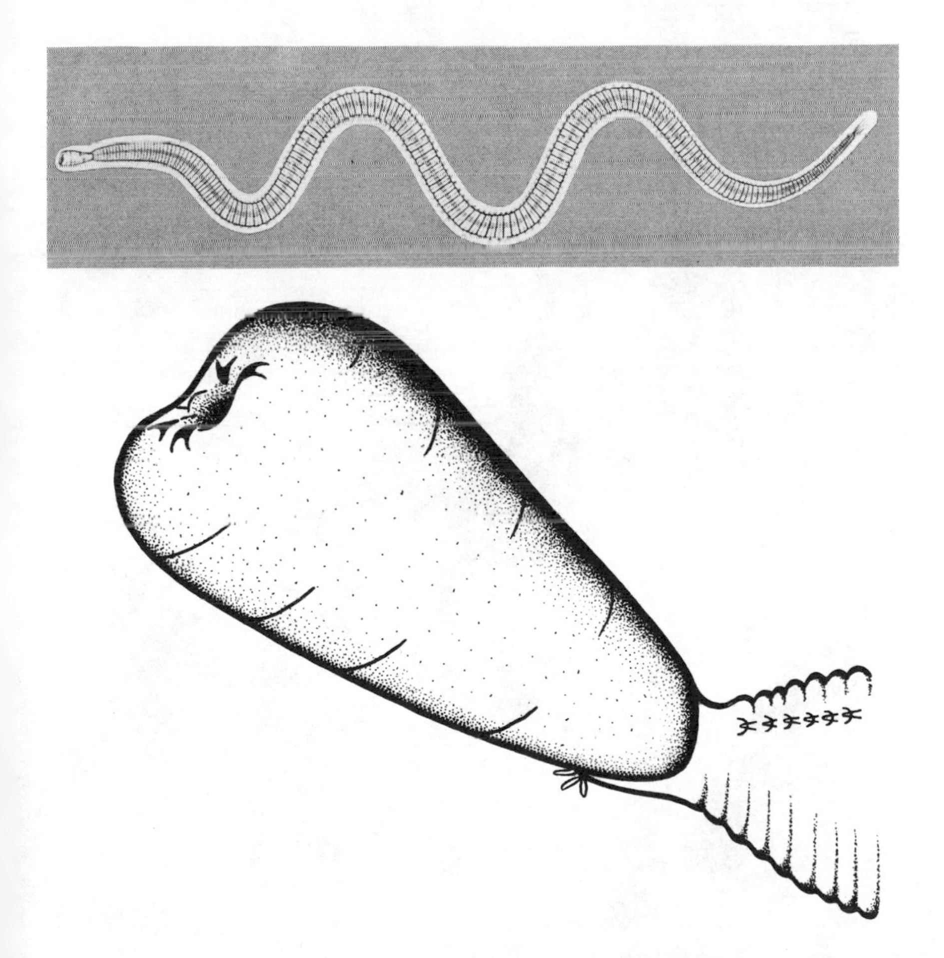

Figure 37. **Bloodworm,** *Glycera dibranchiata* Ehlers, with head enlarged.

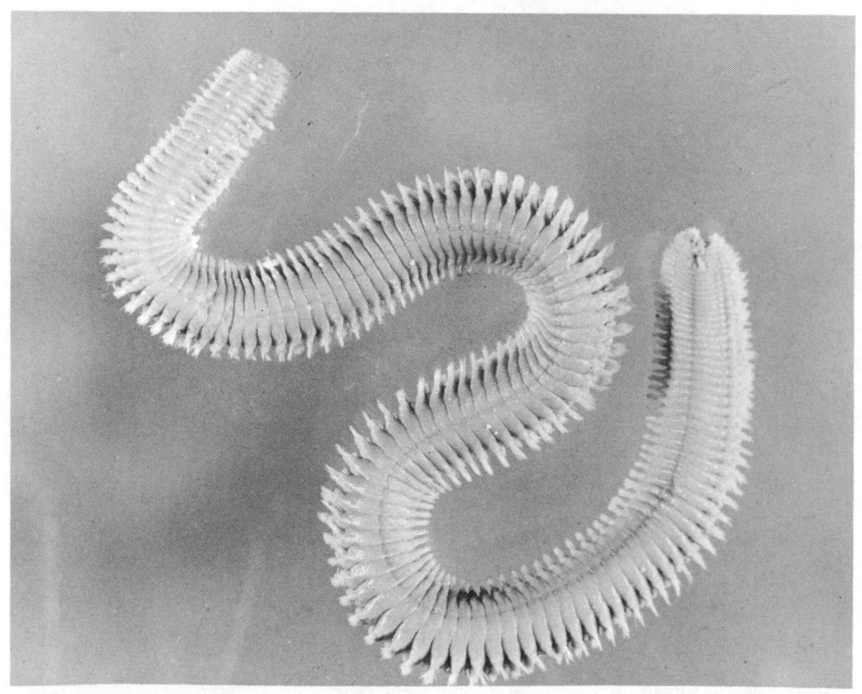
A

B

tlelike structures. In some species these setae are developed into a stinging mechanism. Still other species have tough chitinous jaws with which they can inflict a painful bite. Annelid worms are usually encountered when turning over rocks or coral boulders.

Species of Annelids. There are many species of annelid worms which might be listed, but the following species are representative of the two types: (1) those having biting jaws (*Glycera, Eunice*), and (2) those having stinging setae (*Eurythöe, Hermodice*).

Bloodworm, *Glycera dibranchiata* Ehlers (Figure 37). Found along the coast of North Carolina to northeast Canada.

Bristleworm, *Eurythöe complanata* (Pallas) (Figure 38). Found in the Gulf of Mexico and throughout the tropical Pacific.

Bristleworm, *Hermodice carunculata* Pallas. Inhabits the Gulf of Mexico.

Biting Reef Worm, *Eunice aphroditois* (Pallas) (Figure 39). Circumtropical in distribution.

Medical Aspects: The chitinous jaws of *Glycera* are able to penetrate the skin and to produce a painful sensation similar to that of a bee sting. The marks from the jaws are oval-shaped, and are about as large as the inner circle of the following letter "O," in the center of which is a small reddish spot usually indicating where the jaws have pierced the skin, and bordered by a surrounding area of blanching. The wounded area may become hot and swollen, and may remain so for a day or two. The swelling may be followed by numbness and itching.

Contact with the setae or bristles of the bristleworms may produce inflammation, intense itching, swelling, and numbness, which may persist for several days. The numbness may persist for weeks. It is not definitely known whether a toxin is actually involved. *Eunice aphroditois* may attain a length of 5 feet (1.5 meters) and is equipped with large chitinous jaws that are capable of inflicting a nasty bite. The wounds may be a few millimeters in diameter, soon becoming swollen, hot, and inflamed. These wounds can become infected.

Treatment: The treatment of bristleworm stings is largely symptomatic. The bristles can best be removed by applying adhesive tape to the bristles, or by rubbing the area with sand. The topical application of alcohol or soothing lotions may be helpful and local anesthetic creams,

Figure 38. A. **Bristleworm,** *Eurythöe complanata* (Pallas). B. Enlarged view of *Eurythöe complanata* showing bristles fully extended. *Hermodice carunculata* closely resembles *Eurythöe complanata.* Length 4 inches (10 centimeters) or more.

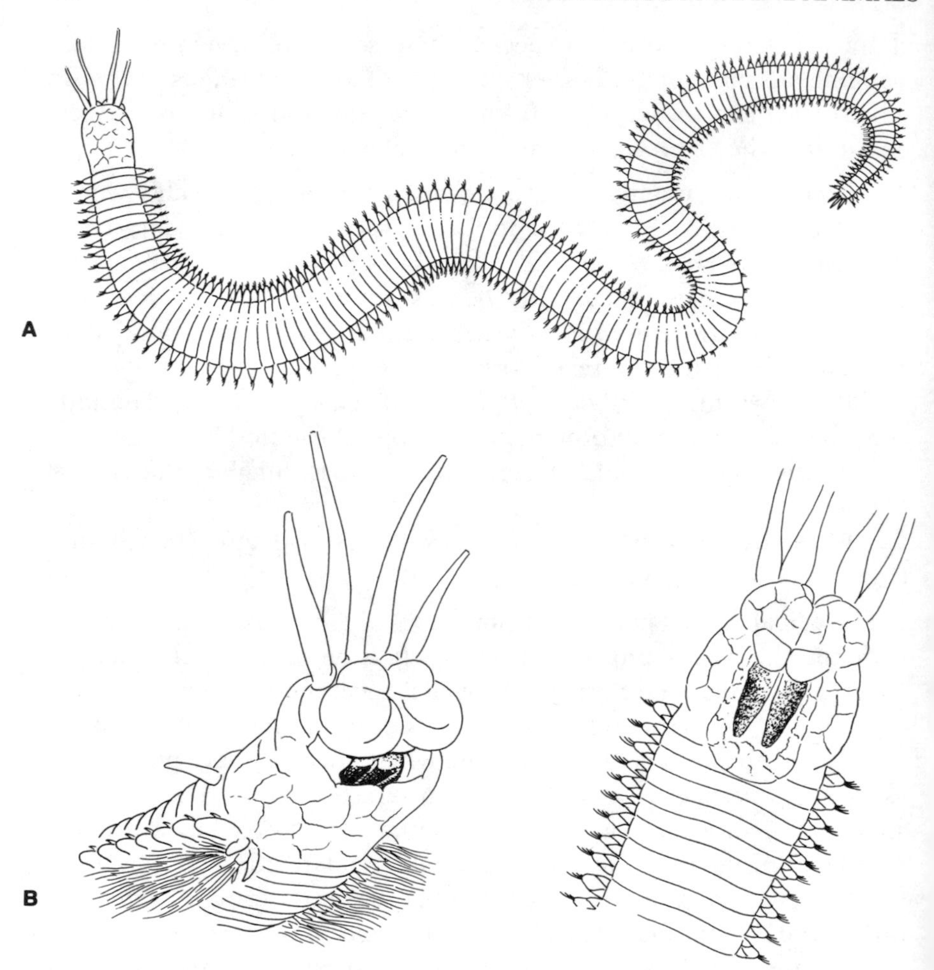

Figure 39. **Biting Reef Worm,** *Eunice aphroditois* (Pallas). A. Adult worm. B. Enlarged view of ventral side of head showing biting jaws. C. Skin removed from lower jaw to reveal biting jaws. This worm is suspected as having venomous jaws and can inflict a painful bite. Attains a length of more than 45 inches (142 centimeters).

ointments, or sprays are useful. Steroid creams, antihistaminics, and antibiotics may be required.

Prevention: Exercise care in the handling of noxious annelid worms. Cotton gloves will probably provide adequate protection from bloodworms, but rubber or heavy leather gloves are advisable when handling bristleworms.

STARFISHES, SEA URCHINS, SEA CUCUMBERS, *ECHINODERMATA*

Echinoderms are characterized by having radial symmetry, with the body usually of five radii around an oral-suboral axis, comprised of calcareous plates which form a more or less rigid skeleton, or with plates and spicules embedded in the body wall. Starfishes or sea stars, sea cucumbers, and sea urchins are all members of this group.

STARFISHES

The Crown of Thorns Starfish, *Acanthaster planci* (Linnaeus) (Figure 40), developed its greatest notoriety because of its massive infestations and destruction of some of the coral reefs in the Indo-Pacific region. The starfish possesses 13 to 16 arms or rays, and may be more than 24 inches (60 centimeters) in diameter. The upper surface of the starfish is covered with numerous long, sharp, stout spines, attaining a length of 2.8 inches (6 centimeters) or more. The spines are covered by a thick integumentary sheath which secretes the venom. These spines can inflict painful wounds. Despite the stout appearance of the spines they are friable and may break off in the wound.

Medical Aspects: Contact with the spines of the venomous starfish may cause extremely painful wounds accompanied by redness, swelling, numbness, and possible paralysis. Nausea and vomiting may be present and the wound may become infected. Some persons are sensitive to the slime of this starfish and contact with the slime may cause a dermatitis.

Treatment: Spines are sometimes embedded and may have to be surgically removed. Exposure of the wound to the suction pads of a living starfish is said to be helpful in alleviating the pain. Immerse the wounded area in hot water (120°F or 50°C), taking care not to scald the patient. If the spine is embedded, a local anesthetic will be required. Steroid creams and antibiotics may be needed. Contact dermatitis is best treated with the use of steroid creams and antihistaminics.

Prevention: Care should be taken in handling this venomous starfish. If possible, wear heavy gloves. Only on the soft undersurface of the animal is it safe to make contact with bare hands.

SEA URCHINS

Some sea urchins are of concern to skin divers because of their abundance and the ability of their sharp spines to penetrate sandals, shoes,

Figure 40. **Crown of Thorns Starfish,** *Acanthaster planci* (Linnaeus). Spines may attain a length of 2.5 inches (6 centimeters) or more, and the starfish more than 24 inches (60 centimeters) in diameter. The spines can inflict a very painful wound. Taken at Heron Island, Great Barrier Reef, Australia.

and flippers. The long-spined tropical sea urchins (*Diadema* and its relatives) are especially hazardous. Persons diving in the West Indies should take special care because of the great abundance of these urchins in shallow water areas in many of the Caribbean Islands. Sea urchins are most commonly encountered under rocks, in crevices, or in other sheltered areas among corals. Sometimes they are found in open sandy flats.

Species of Sea Urchins. Because of the large number of sea urchins, only a few of the more representative ones will be listed.

Long-spined or Black Sea Urchin, *Diadema setosum* (Leske) (Figure 41). Widely distributed throughout the Indo-Pacific area, from East Africa to Polynesia, China, and Japan. Closely related species are found in the West Indies and the Hawaiian Islands. The spines are dangerous.

Figure 41. **Long-spined Sea Urchin,** *Diadema setosum* (Leske). Spines may reach a length of more than 12 inches (30 centimeters) or more Heron Island, Great Barrier Reef, Australia.

Figure 42. **Venomous Sea Urchin,** *Toxopneustes elegans* Doderlein. The venomous globiferous pedicellariae appear as small flowerlike structures extending slightly beyond the spines. The pedicellariae of this species can inflict a painful sting which may be lethal. Do not handle this species without heavy gloves. Taken from Tokyo Bay, Japan.

Sea Urchin, *Toxopneustes pileolus* (Lamarck) Inhabits the Indo-Pacific area, from East Africa to Melanesia, and Japan. Possesses venomous pedicellariae.

Sea Urchin, *Toxopneustes elegans* Doderlein (Figure 42). Inhabits the waters about Japan. Possesses venomous pedicellariae.

Sea Urchin, *Asthenosoma ijimai* Yoshiwara (Figure 43). Ranges from Southern Japan to the Molucca Sea. Possesses sharp venomous spines on the underside of shell.

Venom Apparatus of Sea Urchins: There are two types of venom organs in sea urchins: (1) venomous spines, and (2) their globiferous pedicellariae. Most of the dangerous species are equipped with either one or the other, but seldom both.

Spines. The spines of sea urchins vary greatly from group to group. In most instances the spines are solid, have blunt, rounded tips, and do not constitute a venom organ. However, some species have long, slender, hollow, sharp spines which are extremely dangerous to handle. The acute tips and the spinules permit ready entrance of the spines deep into the flesh, but because of their extreme brittleness the spinules break off

Figure 43. **Sea Urchin,** *Asthenosoma ijimai* Yoshiwara. Part of the outer covering has been removed to show the test beneath. Diameter 6 inches (15 centimeters). Sea Urchin possesses sharp venomous spines on the under surface.

readily in the wound and are very difficult to withdraw. The spines in *Diadema* may attain a length of a foot or more. It is believed that the spines of some of these species secrete a venom, but this has not been experimentally demonstrated. The aboral spines of *Asthenosoma* are developed into special venom organs carrying a single large gland. The point is sharp and serves as a means of introducing the venom (Figure 44).

Pedicellariae. Pedicellariae are small, delicate, seizing organs which are found scattered among the spines of the shell. There are several different types of pedicellariae. One of these, because of its globe-shaped head, is called the globiferous pedicellariae, and serves as a venom organ. The pedicellariae are comprised of two parts, a terminal, swollen, conical head, which is armed with a set of calcareous pincer-like valves or jaws, and a supporting stalk (Figure 45). The head is attached to the stalk either directly by the muscles, or by a long flexible neck. On the inner side of each valve is found a small elevation provided with fine sensory hairs. Contact with these sensory hairs causes the valves to close instantly. The outer surface of each valve is covered by a large gland which in *Toxopneustes* has two ducts that empty in the vicinity of a small toothlike projection on the terminal fang of the valve. A sensory bristle is located on the inside of each valve. Contact with these bristles causes the small muscles at the base of the valve to contract, thus closing the valves and injecting the venom into the skin of the victim.

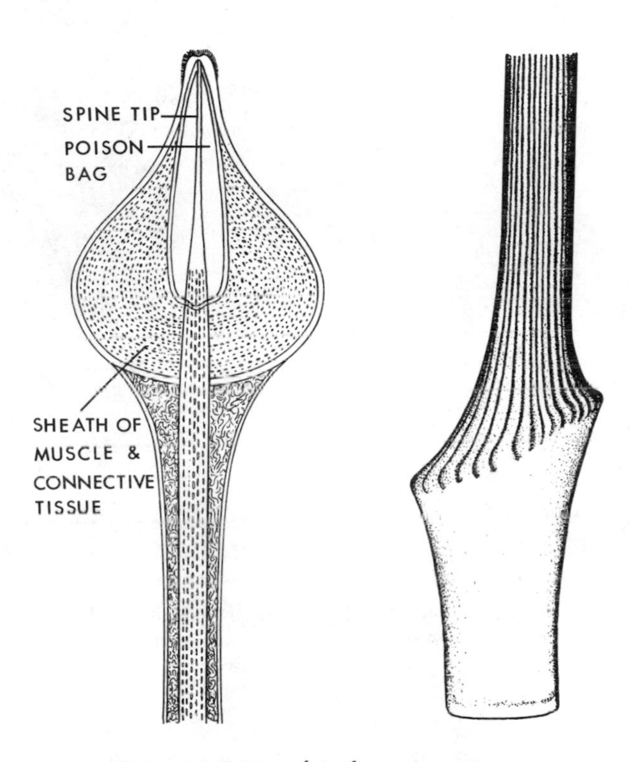

Figure 44. Spines of *Asthenosoma ijimai.*

One of the primary functions of pedicellariae is that of defense. When the sea urchin is at rest in calm water, the valves are generally extended, moving slowly about, awaiting prey. When a foreign body comes in contact with them, it is immediately seized. The pedicellariae do not release their hold as long as the object moves, and if it is too strong to be held, the pedicellariae are torn from the test, or shell, but continue to bite the object. Detached pedicellariae may remain alive for several hours after being removed from the sea urchin, and may continue to secrete venom into the victim.

Medical Aspects: Penetration of the needle-sharp sea urchin spines may produce an immediate and intense burning sensation. The pain is soon followed by redness, swelling, and an aching sensation. Numbness and muscular paralysis have been reported and secondary infections are not uncommon.

The sting from sea urchin pedicellariae may produce an immediate, intense, radiating pain, faintness, numbness, generalized muscular pa-

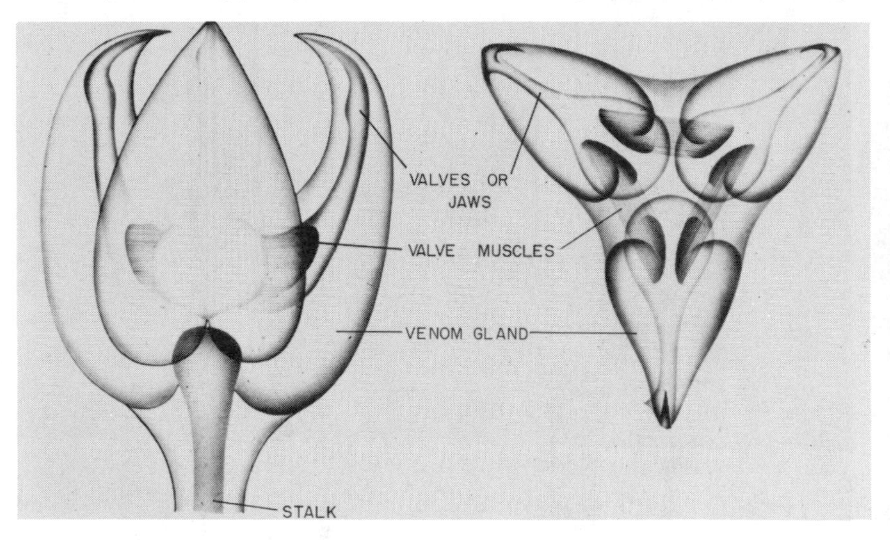

Figure 45. Globiferous pedicellariae from the **Sea Urchin,** *Salamacis bicolor* (Agassiz). This is representative of the venomous type of pedicellariae which are found in other species of sea urchins.

ralysis, loss of speech, respiratory distress, and in severe cases, death. The pain may diminish after about fifteen minutes and completely disappear with an hour, but paralysis may continue for six hours or longer.

Treatment: The treatment of sea urchin wounds caused by their spines varies with the type of spine and the area of the body involved. Most sea urchin spines are friable and break off quite easily when grasped with a pair of forceps. Generally it is difficult to remove sea urchin spines without surgical intervention. If the spine is situated in soft tissue away from bone, it sometimes can be crushed by pummeling the area with the fist. Local anesthesia may be required for pain. This tends to fragment the spine and usually the pieces will be absorbed in a few days and disappear. In other cases, the spines may become encrusted and remain for many months, sometimes migrating to another site where they may have to be surgically removed. Sea urchin spines penetrating a bone joint should be X-rayed, and surgically removed. Failure to remove them may result in a severe chronic inflammatory reaction. Surgical debridement and careful cleansing of the wound in order to remove all of the spine fragments will be needed. If there is pain or instability of the joint, splinting may be necessary. Since sea urchin wounds frequently become infected, hot soaks and antibiotic therapy may be required. Poultices may be useful.

Pedicellariae envenomations may produce severe reactions because they are more potent than the venom found in the spines of sea urchins. Any pedicellariae clinging to the skin should be promptly removed. Bathing the wound with alcohol or some other antiseptic solution is usually helpful. The affected part should be immersed in hot water (120°F or 50°C), being careful not to scald the patient. Artificial respiration may be required. Any allergic reactions can usually be controlled with the use of adrenalin and antihistaminics. Antibiotics may be required. There are no specific antivenins available for sea urchin stings.

Prevention: No sea urchin having elongated, needlelike spines should be handled. Moreover, leather and canvas gloves, shoes, and flippers do not afford protection. Care should be taken in handling any tropical species of short-spined sea urchin without gloves because of the pedicellariae. Because of the danger of coming in contact with sea urchins, a diver working at night in coral areas must exercise extreme care.

IV

MARINE ANIMALS THAT STING: VERTEBRATES

Venomous marine vertebrates constitute a hazard to skin divers which, for the most part, can be easily avoided if the diver is aware of them and knows something about their habits. Included in this chapter are those marine animals having a backbone, which inflict injury by stinging.

Figure 46. **Spiny Dogfish,** *Squalus acanthias* Linnaeus. Note the single spine in front of each of the dorsal fins.

Venomous marine vertebrates fall within two major groups: (1) fishes, and (2) reptiles. A remarkable and interesting variety of venomous fishes is found to exist, beginning with some of the lower forms and extending throughout the piscine group. Venomous marine reptiles, on the other hand, are limited to a relatively few species—the sea snakes, all of which have a similar type of venom apparatus.

FISHES: HORNED OR SPINY SHARKS

There are several species of horned sharks, but only one that has been incriminated to any extent as producing stings in humans. It is the spiny dogfish, *Squalus acanthias* Linnaeus (Figure 46), which is found on both sides of the North Atlantic and North Pacific Oceans. Some of its close relatives are widely distributed throughout temperate and tropical seas.

Dogfish are somewhat sluggish in their movements, and seem to prefer shallow, protected bays for their habitat.

Venom Apparatus. Wounds of the spiny dogfish are inflicted by the dorsal sting which is located adjacent to the anterior margin of each of the two dorsal fins. The venom gland appears as a glistening, whitish substance situated in a shallow groove on the back of the upper portion of each spine. When the spine enters the skin, the venom gland is damaged and the venom enters the flesh of the victim.

Medical Aspects: Symptoms consist of immediate, intense, stabbing pain, which may continue for a period of hours. The pain may be accompanied and followed by redness and severe swelling of the affected part. Tenderness about the wound may continue for several days. According to some authors, dogfish stings have been known to be fatal.

Treatment: Treat as any other fish sting.

Prevention: Stings usually occur from the careless handling of dogfish. Be careful of the dorsal stings when removing the dogfish from a spear, hook, or net. Dogfish can give a sudden jerk and drive the sting deep into the flesh of a reckless fisherman.

STINGRAYS

Stingrays constitute one of the largest and most important groups of venomous marine organisms. There are said to be about 1,500 stingray attacks reported in the United States each year. Stingrays are divided into seven families:

1. *Dasyatidae*—stingrays or whiprays
2. *Potamotrygonidae*—river rays
3. *Gymnuridae*—butterfly rays
4. *Myliobatidae*—eagle rays or bat rays
5. *Rhinopteridae*—cow-nosed rays
6. *Mobulidae*—devil rays or mantas
7. *Urolophidae*—round stingrays

The species within these families are too numerous to list individually, so only a few representative species, which have been studied to a considerable extent by the venomologist, will be included. However, there are a few generalities which are pertinent to the group as a whole.

Rays are common inhabitants of tropical, subtropical, and warm temperate seas (Figure 47). With the exception of the family *Potamotrygonidae*, which is confined to freshwater, rays are essentially marine forms, some of which may enter brackish, or freshwaters, freely. Although most common in shallow water, rays swim in moderate depths as well. A deep sea species has been reported from the Central Pacific Ocean. Sheltered bays, shoal lagoons, river mouths, and sandy areas

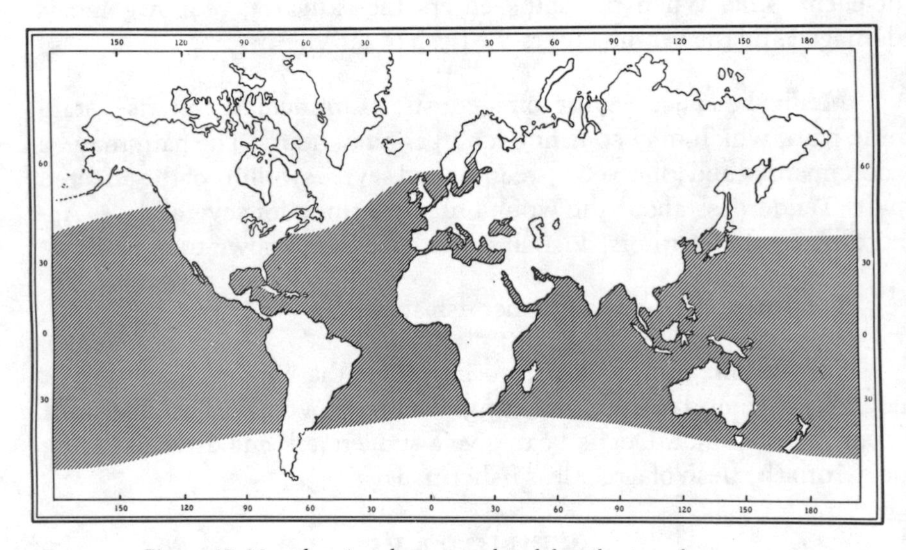

Figure 47. Map showing the geographical distribution of stingrays.

between patch reefs are favorite habitats of rays. They may be observed lying on top of the sand, or partially submerged, with only their eyes, spiracles, and a portion of the tail exposed. Rays burrow into the sand and mud, excavating with the use of their pectoral fins worms, molluscs, and crustaceans upon which they feed.

REPRESENTATIVE STINGRAY SPECIES

STINGRAYS OR WHIPRAYS, *DASYATIDAE*

Giant Stingray of Australia, *Dasyatis brevicaudata* (Hutton). Inhabits the Indo-Pacific region (Figure 48). Reputed to be the largest stingray in the world, attaining a length of 15 feet (4.5 meters), a width of 7 feet (2.2 meters), and a weight of more than 715 pounds (325 kilograms).

Diamond Stingray, *Dasyatis dipterurus* (Jordan and Gilbert) (Figure 49). Occurs from British Columbia to Central America.

European Stingray, *Dasyatis pastinaca* (Linnaeus) (Figure 50). Inhabits the northeastern Atlantic Ocean, Mediterranean Sea, and Indian Ocean. It is a very common species.

Bluntnose Stingray, *Dasyatis sayi* (Le Sueur) (Figure 51). Inhabits the Western Atlantic from New Jersey to southern Brazil.

Figure 48. **Giant Stingray,** *Dasyatis brevicaudata* (Hutton). Reputed to be the largest stingray in the world, this specimen was estimated to have a width of about 5 feet (1.5 meters). Taken near Heron Island, Great Barrier Reef, Australia.

BUTTERFLY RAYS, *GYMNURIDAE*

Butterfly Ray, *Gymnura marmorata* (Cooper) (Figure 49). Found from Point Conception, California, south to Mazatlan, Mexico.

BAT RAYS OR EAGLE RAYS, *MYLIOBATIDAE*

Bat Ray or **Spotted Eagle Ray,** *Aetobatus narinari* (Euphrasen) (Figure 52). Inhabits tropical and warm-temperate belts of the Atlantic, Red Sea, and Indo-Pacific Oceans.

California Bat Ray, *Myliobatis californicus* (Gill) (Figure 52). Found from Oregon to Magdalena Bay, Lower California.

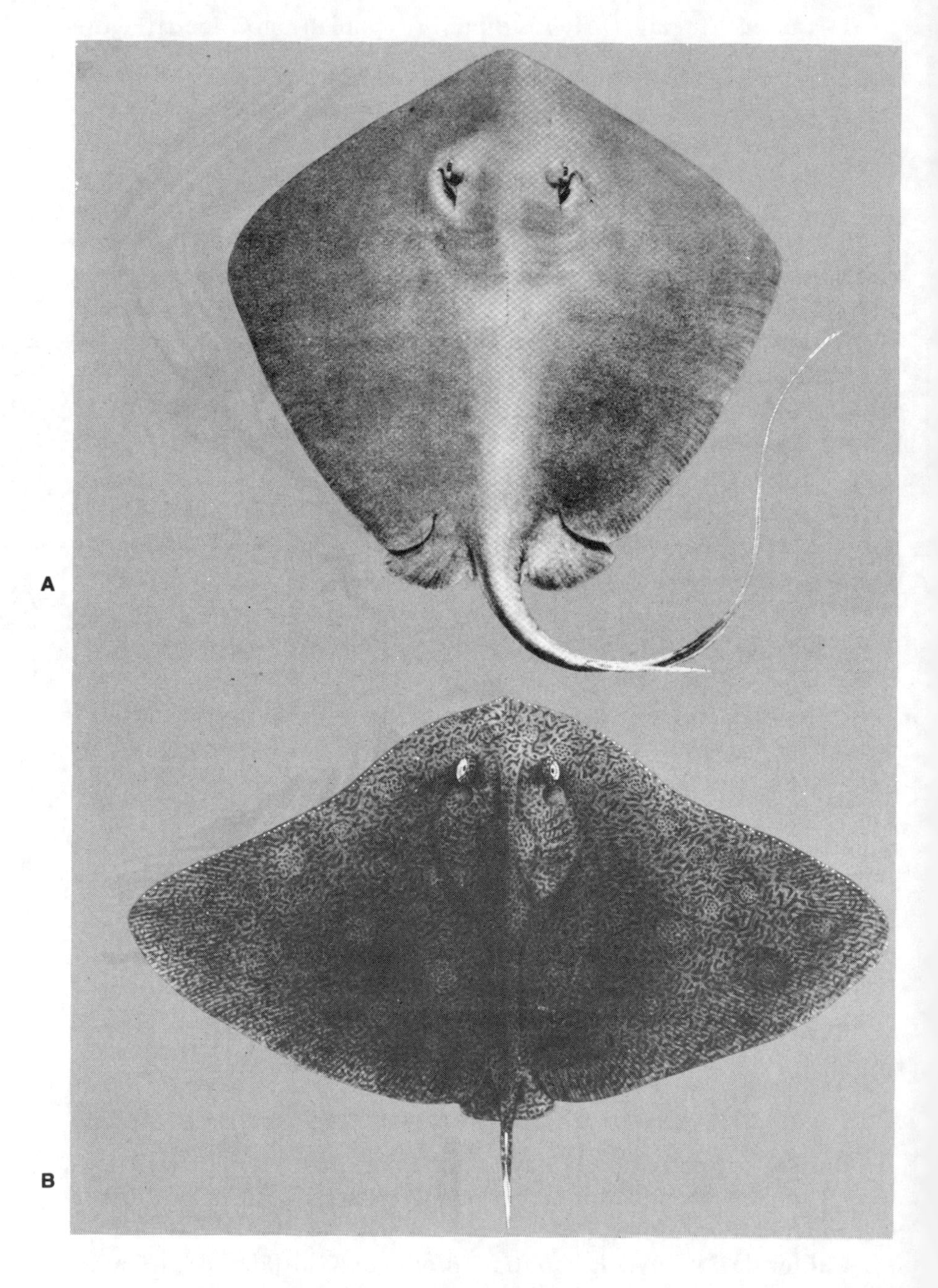

Figure 49. A. **Diamond Stingray,** *Dasyatis dipterurus* (Jordan and Gilbert). B. **Butterfly Ray,** *Gymnura marmorata* (Cooper). (From Hiyama)

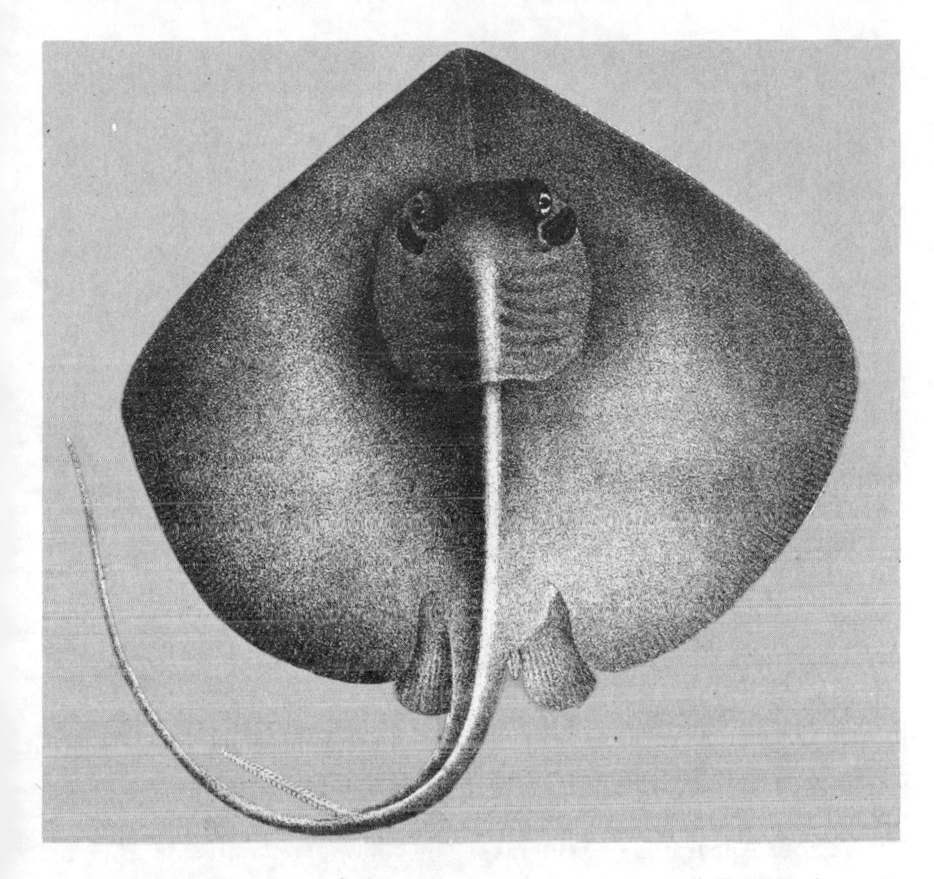

Figure 50. **European Stingray,** *Dasyatis pastinaca* (Linnaeus). (From Day)

FRESHWATER STINGRAYS, *POTAMOTRYGONIDAE*

South American Freshwater Stingray, *Potamotrygon motoro* (Müller and Henle) (Figure 53). Found in the freshwater rivers of Paraguay, and the Amazon River, south to Rio de Janeiro, Brazil. It is an extremely dangerous species.

COW-NOSED RAYS, *RHINOPTERIDAE*

Cow-nose Ray, *Rhinoptera bonasus* (Mitchill) (Figure 54). Found in the coastal western Atlantic, from New England to Brazil.

ROUND STINGRAYS, *UROLOPHIDAE*

Round Stingray, *Urolophus halleri* Cooper (Figure 53). Found from Point Conception, California, south to Panama Bay.

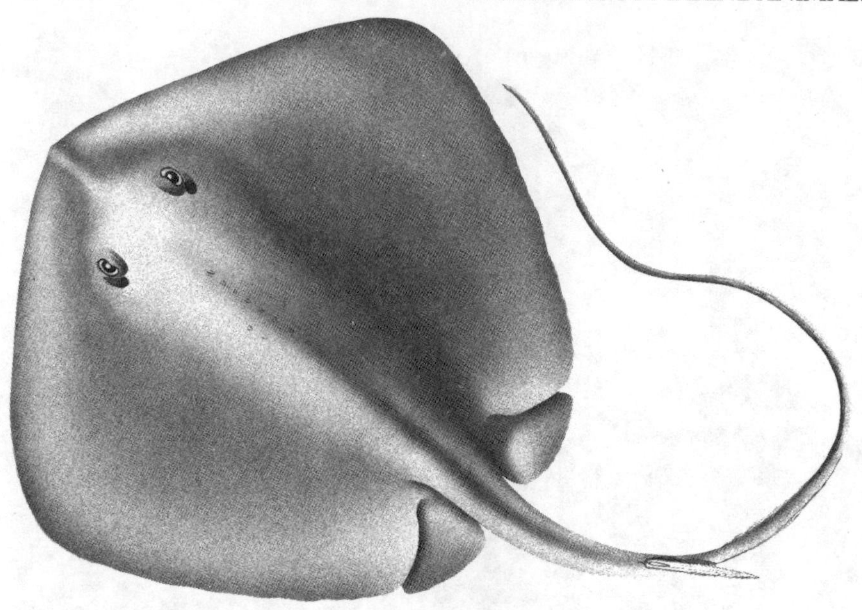

Figure 51. **Bluntnose Stingray,** *Dasyatis sayi* (Le Sueur).

Yellow Stingray, *Urolophus jamaicensis* (Cuvier) (Figure 55). Found in the western tropical Atlantic, from Florida southward into the Caribbean.

Venom Apparatus of Stingrays: The venom apparatus, or sting, of stingrays is an integral part of the caudal, or tail, appendage. A study of stingrays reveals that there are four general anatomical types of venom organs which vary somewhat in their effectiveness as a defensive weapon. The types are shown in Figure 56.

1. *Gymnurid type.* This type is found in the butterfly rays (*Gymnura*). The sting is small, poorly developed, and situated close to the base of a short tail, making it a relatively feeble striking organ.

2. *Myliobatid type.* This type is found in the bat, or eagle rays (*Myliobatis, Aetobatis, Rhinoptera*). Their tails terminate in a long whiplike appendage. The stings in these rays are frequently large and well-developed, but situated near the base of the tail. Under the proper circumstances these rays can use their venom organs to good advantage.

3. *Dasyatid type.* This type is found in the stingrays, proper

Figure 52. A. **Spotted Eagle Ray,** *Aetobatus narinari* (Euphrasen). B. **California Bat Ray,** *Myliobatis californicus* (Gill). ▶

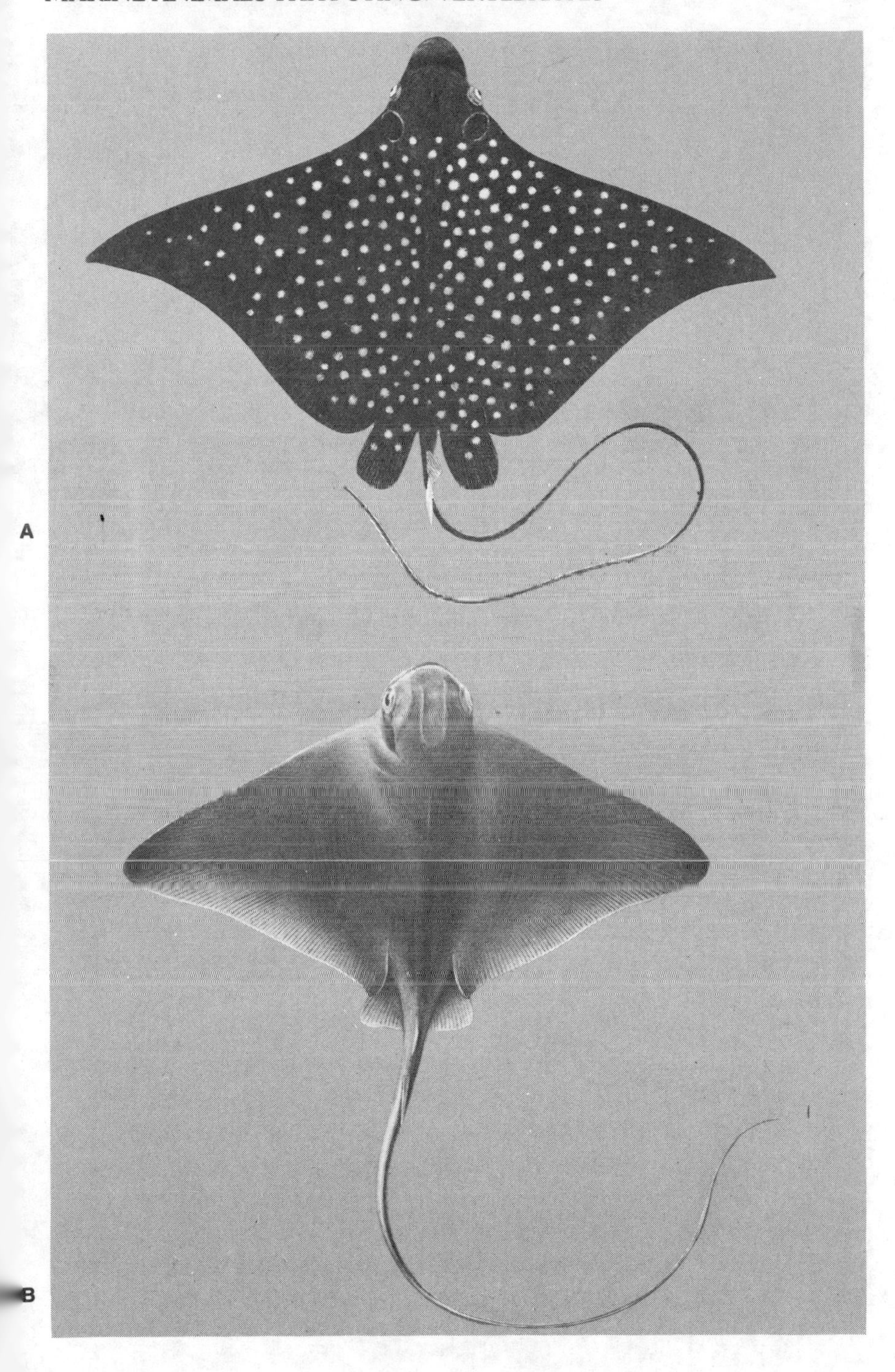
A
B

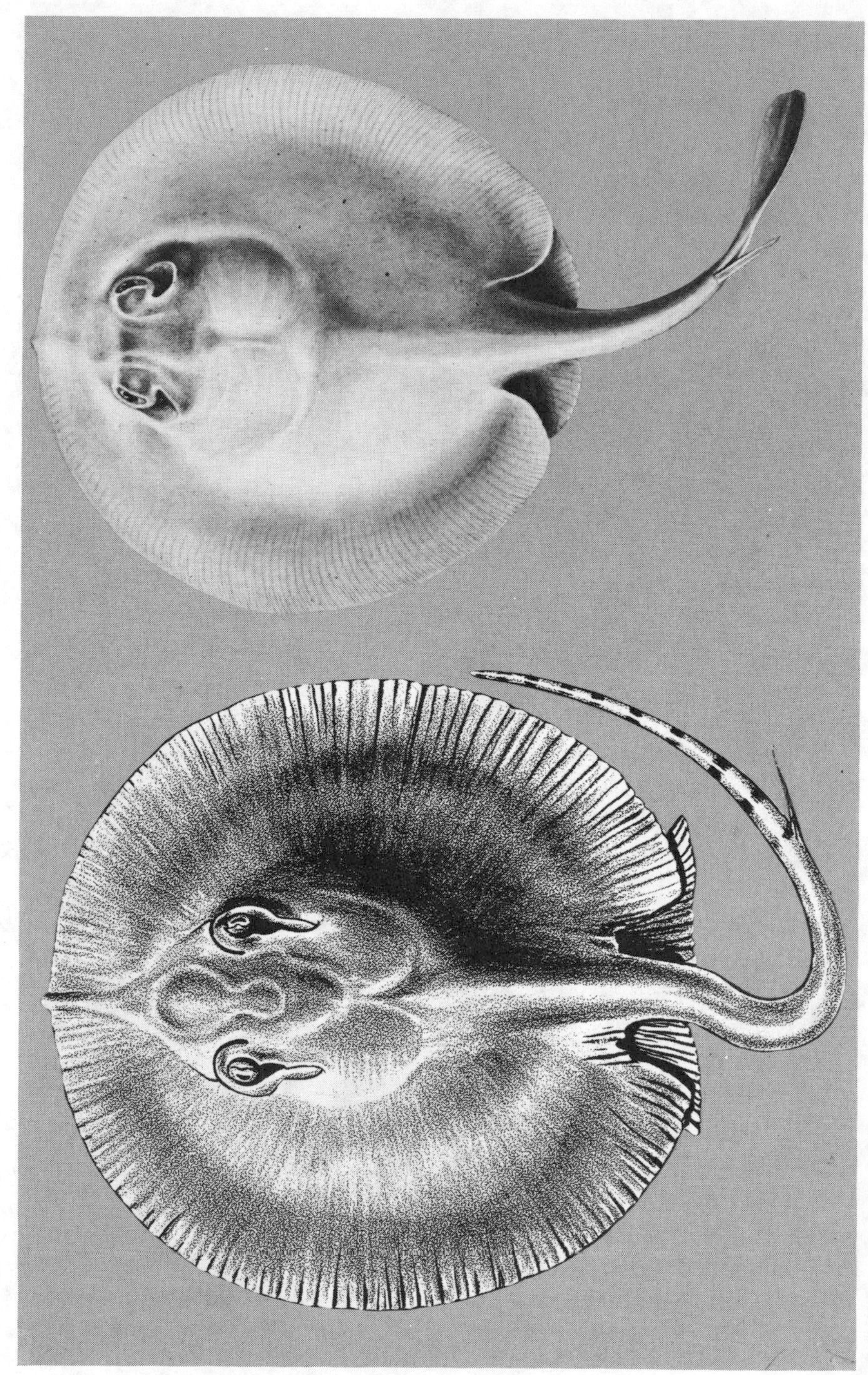
A
B

Figure 54. **Cow-nose Rays,** *Rhinoptera bonasus* (Mitchill).

(*Dasyatis* and *Potamotrygon*). The sting is well-developed as in the previous type, but is located further out from the base of the tail, making it a more effective striking organ. Stingrays possessing this type of venom apparatus are among the most dangerous kinds known. Their tails terminate in a long whiplike appendage.

4. *Urolophid type.* This type is found among the round stingrays (*Urolophus*). The caudal appendage to which the sting is attached is short, muscular, and well-developed. Urolophid rays are also dangerous to man, and can inflict a well-directed sting.

In general, the venom apparatus of stingrays consists of the serrate spine and an enveloping sheath of skin. Together they are termed the sting. Stingrays usually possess only a single spine at a time, but it is not unusual to find a specimen with two or more. Apparently, the spine remains until it is removed by injury. There is no evidence to support the idea that the spines are shed each year. As the young spine grows out from the tail, it takes with it a layer of skin, the so-called integumentary sheath, which continues to ensheathe the spine until it is removed by injury or wear.

The spine is composed of a hard, bonelike material, called vasodentine. Along both sides of the spine are a series of sharp recurved

◂ Figure 53. A. **South American Freshwater Stingray,** *Potamotrygon motoro* (Müller and Henle). (After Riberio) B. **Round Stingray,** *Urolophus halleri* Cooper.

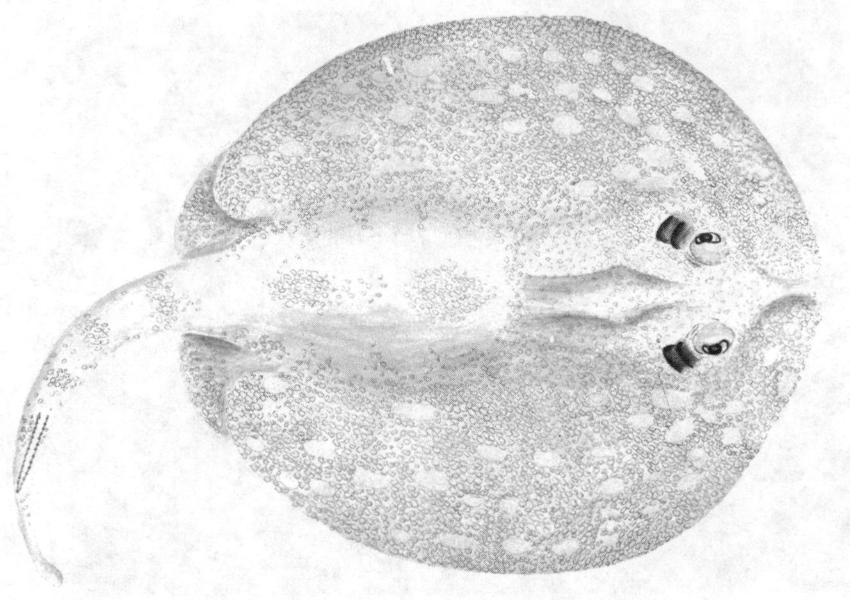

Figure 55. **Yellow Stingray,** *Urolophus jamaicensis* (Cuvier).

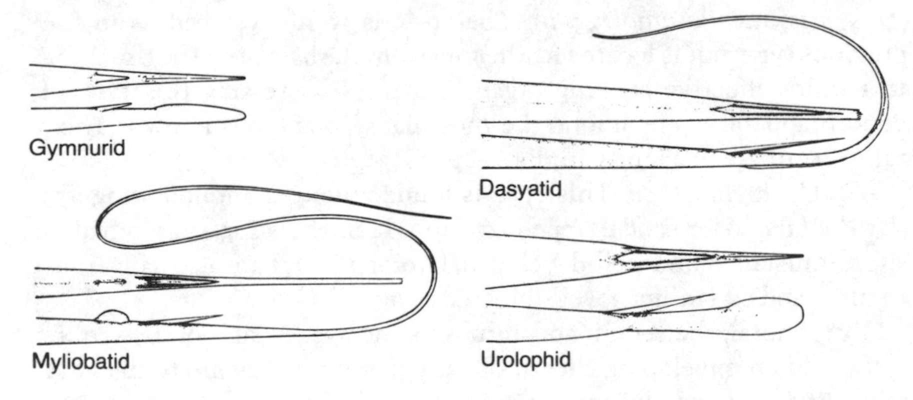

Figure 56. Anatomical types of venom organs that are found in the various species of stingrays. (From Halstead and Bunker)

teeth. The spine is marked by a number of irregular, shallow furrows which run almost the length of the spine. Along either edge, on the underside of the spine, there will be found deep grooves; these grooves are technically termed the ventrolateral-glandular grooves. If these grooves are carefully examined, it will be observed that they contain a strip of soft, spongy, grayish tissue extending throughout their length. The bulk of the venom is produced by this tissue in the grooves, although lesser amounts are believed to be produced by other portions of the integumentary sheath and in certain specialized areas of the skin on

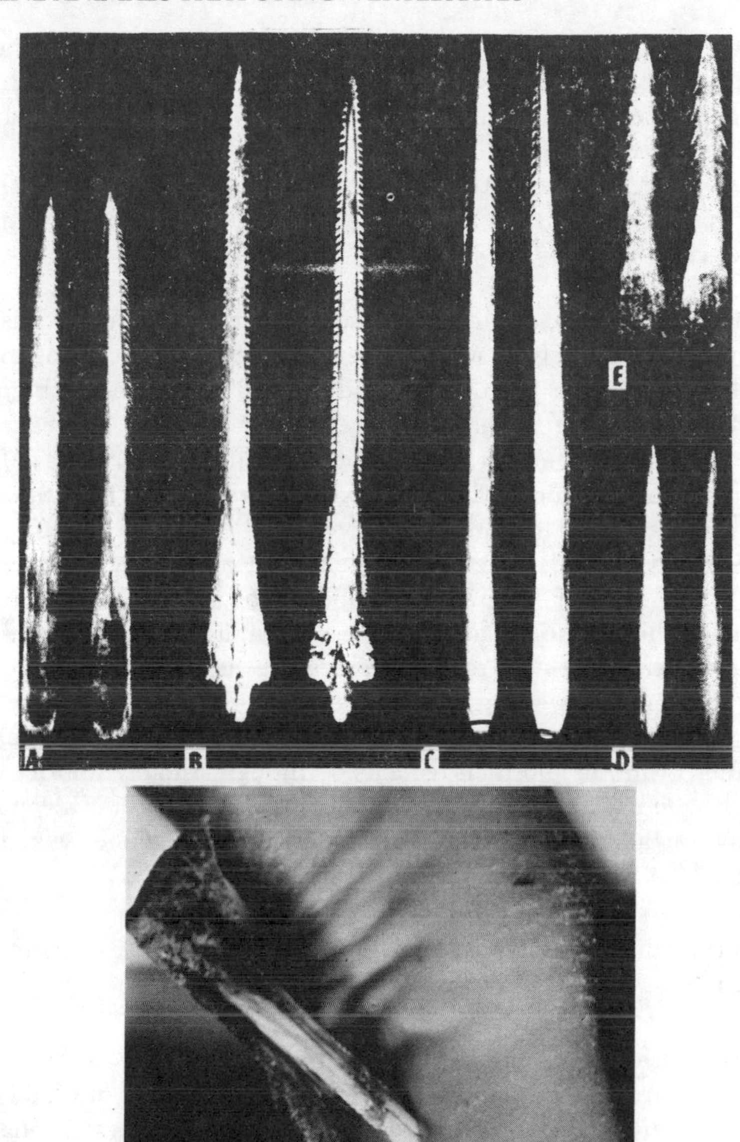

Figure 57. A. Caudal stings from various species of stingrays. a. *Myliobatis californicus.* b. *Aetobatus narinari.* c. *Dasyatis dipterurus.* d. *Urolophus halleri.* e. *Gymnura marmorata.* Dorsal and ventral views. All are mature spines except for *Gymnura marmorata.* B. Heel injury caused by a boy stepping on a dead stingray. (Courtesy Robert D. Hayes)

the tail adjacent to the spine. These grooves serve to protect the soft delicate glandular tissue which lies within them, and even though all of the integumentary sheath may be worn away, the venom-producing tissue continues to remain within these grooves. Thus, a perfectly clean-looking spine can still be venomous. Plate 7 illustrates the sting-ray sting in cross section. Figure 57 shows the caudal spines of various species of stingrays.

Medical Aspects: Pain is the predominant symptom and usually develops immediately or within a period of ten minutes following the attack. The pain has been variously described as sharp, shooting, spasmodic or throbbing in character. Freshwater stingrays are reputed to cause extremely painful wounds. More generalized symptoms of fall in blood pressure, vomiting, diarrhea, sweating, rapid heart beat, muscular paralysis, and death have also been reported.

Stingray wounds are either of the laceration or puncture type (Figure 57). Penetration of the skin and underlying tissue is usually accomplished without serious damage to the surrounding structures, but withdrawal of the sting may result in extensive tissue damage due to the recurved spines. Swelling in the vicinity of the wound is a constant finding. The area about the wound at first has an ashy appearance, later becomes cyanotic and then reddens. Although stingray injuries occur most frequently about the ankle joint and foot as a result of stepping on the ray, instances have been reported in which the wounds were in the chest. Chest wounds may result in death.

Treatment: There is no known specific antidote. See treatment of fish stings.

Prevention: It should be kept in mind that stingrays commonly lie almost completely buried in the upper layer of a sandy or muddy bottom. They are, therefore, a hazard to anyone wading in water inhabited by them. The chief danger is in stepping on one that is buried (Figure 58). Since the body of the stingray is usually pinned down by the weight of the victim, thereby permitting the beast to make a successful strike, pushing or shuffling one's feet along the bottom eliminates the danger and at the same time chases the stingray from its lair. It is also recommended that a stick be used to probe along the bottom in order to rid the area of hidden rays.

ELEPHANTFISHES AND RATFISHES

Chimaeroids, elephantfish, or ratfish, as they are variously called, are a group of cartilaginous fishes having a single external gill opening on either side, covered over by a skin fold which leads to the gill chamber. Externally, ratfishes are more or less compressed laterally, tapering posteriorly to a slender tail. The snout is rounded, or cone-shaped, extended as a long pointed beak, or bears a curious hoe-shaped probos-

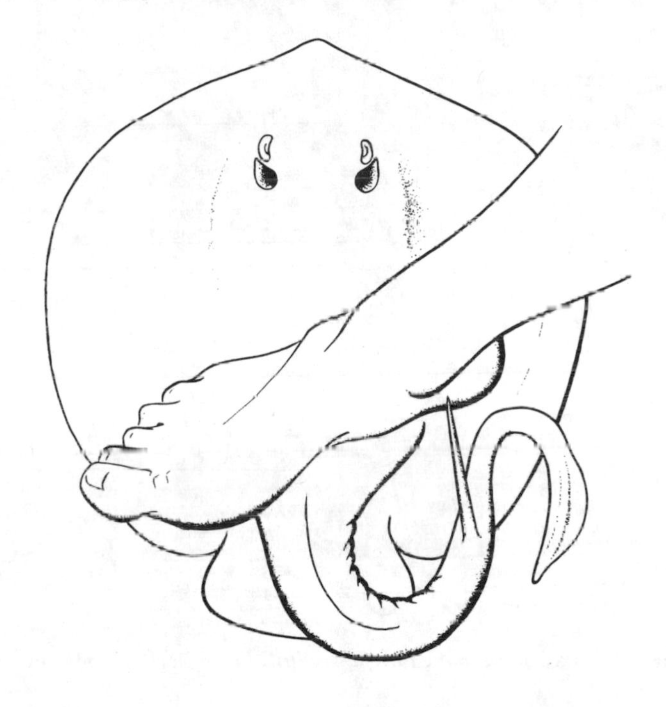

Figure 58. Drawing showing how stingray wounds are most frequently encountered.

cis. There are two dorsal fins. The first fin is triangular, usually higher than the second, and edged anteriorly by a strong, sharp-pointed bony spine which serves as a venom organ (Figure 59). Ratfishes have a preference for cooler waters, and have a depth range from the surface down to 1,400 fathoms (2,625 meters). They are weak swimmers and die soon after being removed from the water. They have well-developed dental plates and can inflict a nasty bite. Only two species have been studied to any extent by venomologists.

SPECIES OF RATFISHES

European Ratfish, *Chimaera monstrosa* Linnaeus (Figure 60). Inhabits the north Atlantic Ocean from Norway and Iceland to Cuba, the Azores, Morocco, Mediterranean Sea, and South Africa.

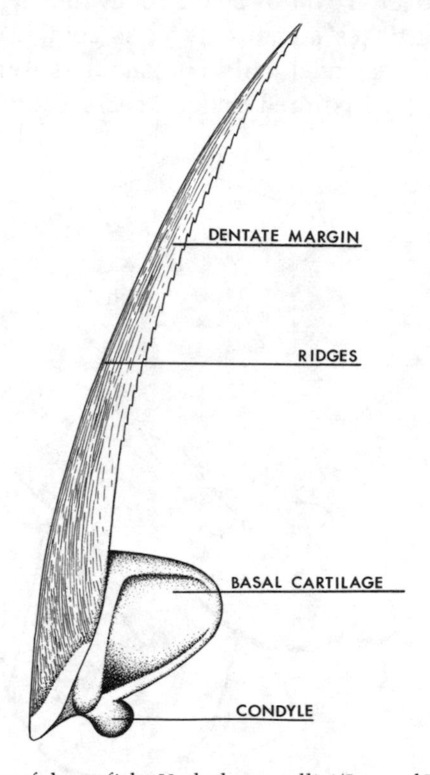

Figure 59. Dorsal spine of the ratfish, *Hydrolagus colliei* (Lay and Bennett).

Pacific Ratfish, *Hydrolagus colliei* (Lay and Bennett) (Figure 60). Found along the Pacific coast of North America.

Venom apparatus of the Ratfish: The venom apparatus of ratfishes consists of the single dorsal sting which is situated along the anterior margin of the first dorsal fin. Along the back of the spine is a shallow depression which contains a strip of soft, grayish tissue—the venom gland.

Medical Aspects: Ratfishes inflict a very painful single puncture wound with their sharp dorsal stings. The pain is immediate and tends

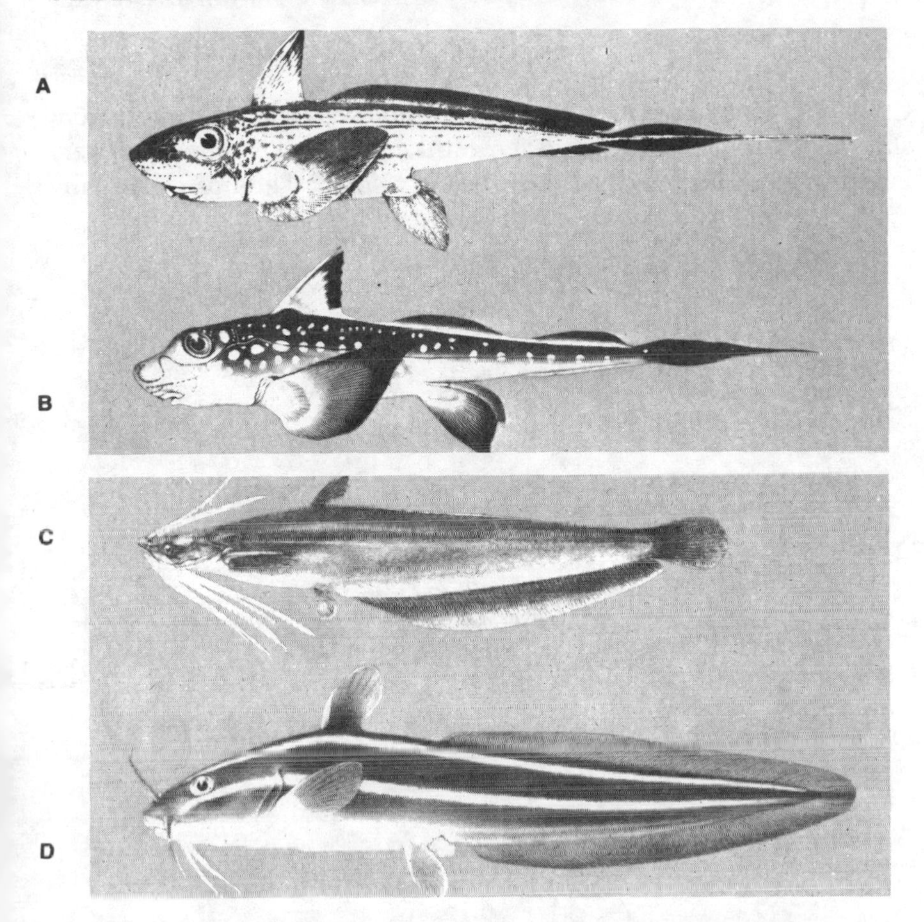

Figure 60. A. **European Ratfish,** *Chimaera monstrosa* Linnaeus. B. **Pacific Ratfish,** *Hydrolagus colliei* (Lay and Bennett). C. **Catfish,** *Heteropneustes fossilis* (Bloch). D. **Oriental Catfish,** *Plotosus lineatus* (Thunberg).

to increase in severity for a few minutes, and may continue to be severe for several hours, gradually lessening in intensity. A dull ache may continue for many days. The area about the wound may become numb to touch and develop a cyanotic or blackish appearance. The outer area may be pale, swollen, and similar in appearance to a severe inflammatory reaction. Joint aches and swollen lymph nodes may be present. It is assumed that wounds from this fish are largely accidental since the fish is not reputed to be aggressive. However, they are also equipped with powerful jaws and are capable of inflicting nasty bites. The viscera of these fish have been found to be toxic to laboratory animals.

CATFISHES

Catfishes come in a wide variety of sizes and body shapes, varying from short to greatly elongated, or even eellike. the head is sometimes very large, wide or depressed, or may be very small. The lips are usually

Figure 61. A. **Catfish,** *Galeichthys felis* (Linnaeus). B. **Catfish,** *Clarias batrachus* (Linnaeus). C. **Sea Catfish,** *Bagre marina* (Mitchill).

equipped with long barbels. The skins of these fishes are thick and slimy, or may bc covered with bony plates. No true scales are ever present. Included within this group are about 1,000 species, most of which are found in the freshwater streams of the tropics. A few are marine. For the purposes of this manual, only a few representative marine and brackish water species have been selected.

REPRESENTATIVE SPECIES OF CATFISHES

Catfish, *Galeichthys felis* (Linnaeus) (Figure 61). Ranges from cape Cod to the Gulf of Mexico.

Catfish, *Clarias batrachus* (Linnaeus) (Figure 61). India to the Netherlands, Indies and Philippine Islands.

Catfish, *Heteropneustes fossilis* (Bloch) (Figure 60). Found along the coasts of India, Ceylon, and Viet Nam.

Oriental Catfish, *Plotosus lineatus* (Thunberg) (Figure 60). Occurs in the vicinity of river mouths throughout much of the Indo-Pacific area.

Sea Catfish, *Bagre marinus* (Mitchill) (Figure 61). Inhabits the east coast of America from Cape Cod to Rio de Janeiro, Brazil.

Venom Apparatus of Catfishes: Venomous catfishes have a single, sharp, stout spine immediately in front of the soft-rayed portion of the dorsal and pectoral fins (Figure 62). This spine is enveloped by a thin layer of skin, the integumentary sheath, which is continuous with that of the soft-rayed portion of the fin. There is no external evidence of the venom glands, which are located as a series of glandular cells within the outer, or epidermal, layer of the integumentary sheath. The venom, or glandular, cells are most concentrated at the anterolateral and posterolateral margins of the sting where they are sometimes clumped two or three cells deep within the epidermal layer. The spines of some species are also equipped with a series of sharp, recurved teeth which are capable of producing a severe laceration of the victim's flesh, thus facilitating absorption of the venom and subsequent secondary infection (Figure 63). The spines of the catfish are particularly dangerous because they can be locked into the extended position at the will of the fish.

Medical Aspects: The pain is generally described as an instantaneous stinging, throbbing, or scalding senation which may be localized or may radiate up the affected limb. Some of the tropical species, such as *Plotosus*, are capable of producing violent pain which may last for 48 hours or more. Immediately after being stung, the area about the wound becomes pale. The pallor is soon followed by a cyanotic appearance, and then by redness and swelling. In some cases swelling may be very severe, accompanied by numbness and gangrene of the area about the wound. Shock may be present. Improperly treated cases frequently result in secondary bacterial infections of the wound. Because of recurved teeth

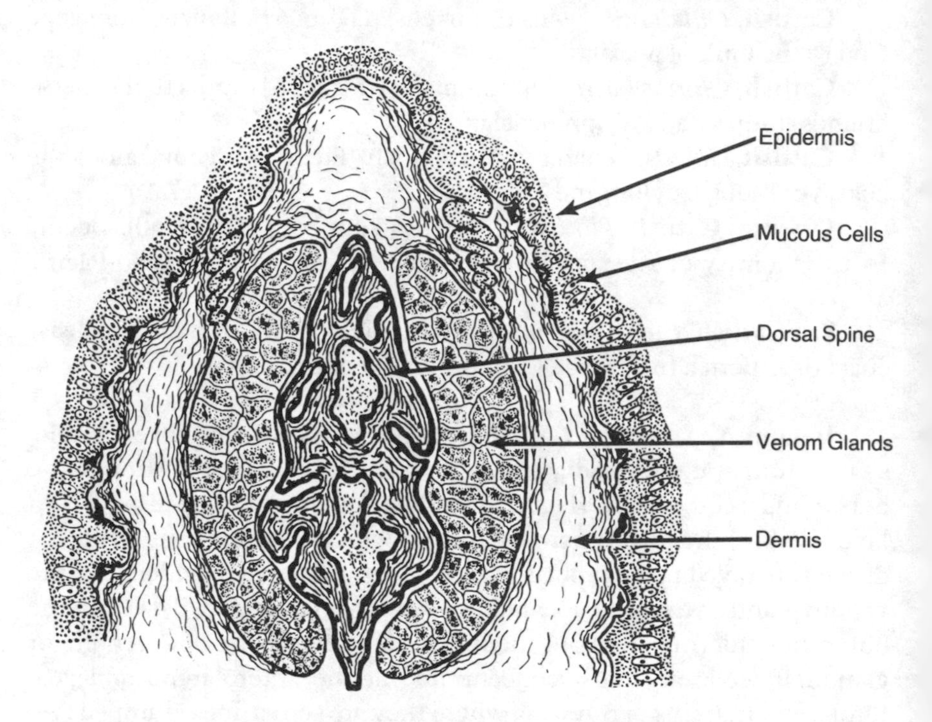

Figure 62. Drawing of a cross section of the dorsal sting of *Plotosus lineatus*. Note the large platelike arrangement of the venom glands which cover most of the lateral surfaces of the dorsal spine.

along the margin of their spines, some species of catfishes may produce wounds which may take weeks to heal. Deaths have been reported from stings of some tropical catfishes.

Treatment: There are no known specific antidotes. See treatment of fish stings.

Prevention: Care should be exercised in the handling of catfishes because of their sharp rigid fin spines which can be readily driven into the flesh of the unwary victim (Figure 64).

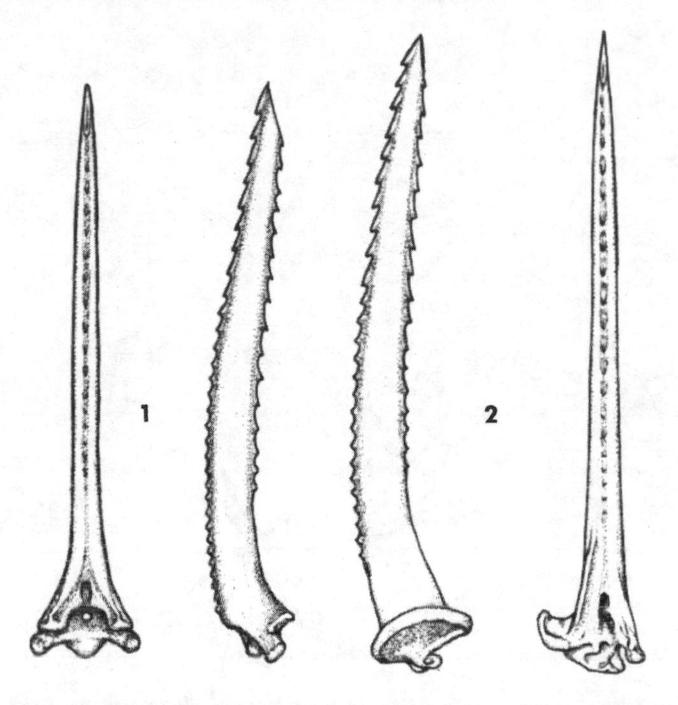

Figure 63. Fin spines of the **Catfish,** *Galeichthys felis.* 1. Dorsal spine, posterior and side views. 2. Pectoral spine, dorsal and posterior views.

MORAY EELS

Moray eels are members of the family *Muraenidae*. The bodies of these fishes are scaleless, elongated, and rounded, or more or less compressed. The dorsal and anal fins are continuous with the caudal, and are generally covered by thick skin. The pectoral and pelvic fins are absent.

The meager amount of research that has been conducted on so-called "venomous eels" is concerned with the single European species, *Muraena helena* Linnaeus, which is distributed along coastal areas of the eastern Atlantic Ocean and Mediterranean Sea. Other species have been listed as venomous by modern authors, but their opinions concerning these species are purely presumptive.

The teeth of *Muraena helena* have been considered as constituting a venom apparatus from earliest times. Anatomical studies fail to show any evidence of venom glands. However, it has been suggested that the

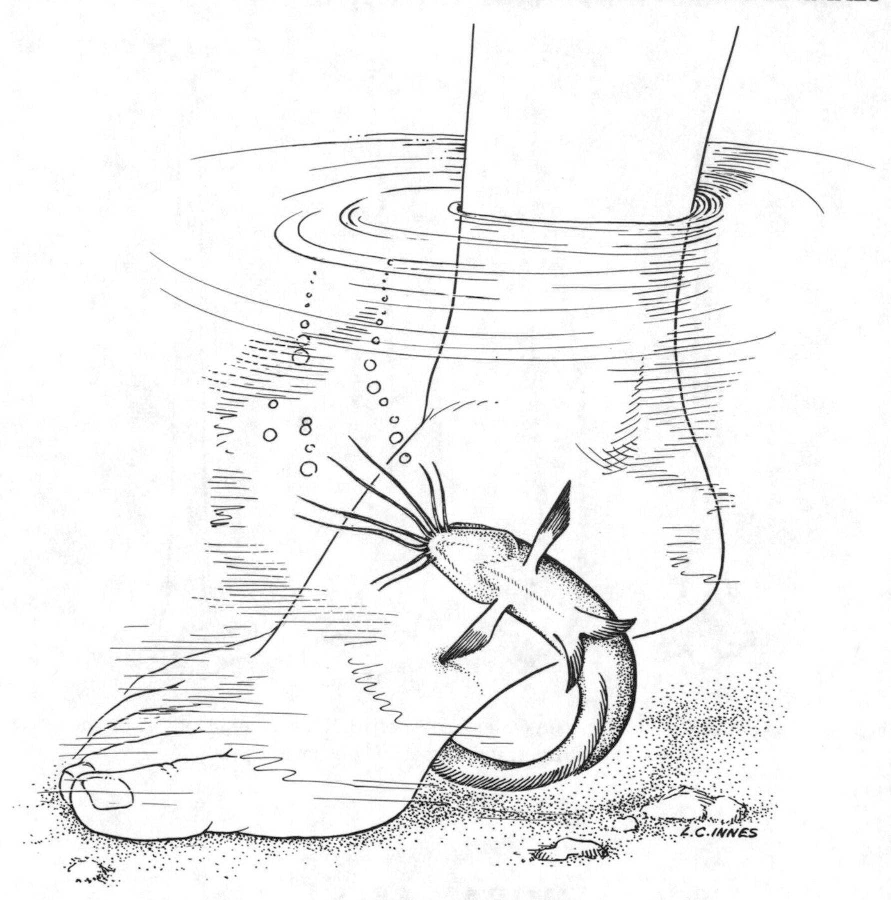

Figure 64. Drawing showing how some catfishes are able to inflict a wound with their dorsal spine.

palatine mucosa may secrete a toxic substance. The bite of a moray can be very painful and may become infected.

Treatment: Treat the same manner as other types of fish stings and wounds.

WEEVERFISHES

All weevers are small marine fishes which attain a maximum length of less than 18 inches (46 centimeters). They are members of the family

Trachinidae, and are among the more venomous fishes of the temperate zone. Primarily dwellers of flat, sandy, or muddy bays, weevers are commonly seen burying themselves in the soft sand or mud with only their heads partially exposed. They may dart out rapidly and may strike an object with their cheek spines with unerring accuracy. When a weever is provoked, the dorsal fin is instantly erected, and the gill covers expanded. Because of their habit of concealment, aggressive attitude, and highly-developed apparatus, they constitute a real danger to any skin diver working in their habitat. There are four species of weevers which are commonly recognized, but only two of them are included here.

SPECIES OF WEEVERFISHES

Greater Weever, *Trachinus draco* Linnaeus (Figure 65). Occurs from Norway, British Isles, southward to the Mediterranean Sea, and along the coast of North Africa.

Lesser Weever, *Trachinus vipera* Cuvier (Figure 65). Inhabits the North Sea, southward along the coast of Europe, and Mediterranean Sea.

Venom Apparatus of Weevers: The venom apparatus of the weeverfish consists of the dorsal and opercular spines and their associated glands (Figure 66). The dorsal spines vary from five to seven in number. Each of the spines is enclosed within a thin-walled sheath of skin from which protrudes a needle-sharp tip (Figure 66). Removal of the sheath reveals a thin, elongated, fusiform strip of whitish spongy tissue lying within the grooves, near the tips, of each spine. This spongy tissue is the venom-producing part of the spine. Removal of the skin covering the gill cover shows a broad, compressed, "daggerlike" opercular spine ending in a sharp tip (Figure 66). Attached to the upper and lower margins of the spine are the pear-shaped venom glands. Weever venom has been found to act as both a neurotoxin and hemotoxin, similar to some snake venoms.

Medical Aspects: Weever wounds usually produce instant pain, described as a burning, stabbing or "crushing" sensation—initially confined to the immediate area of the wound, then gradually spreading through the affected limb. The pain gets progressively worse until it reaches an excruciating peak, generally within 30 minutes. The severity is such that the victim may scream, thrash wildly about, and lose

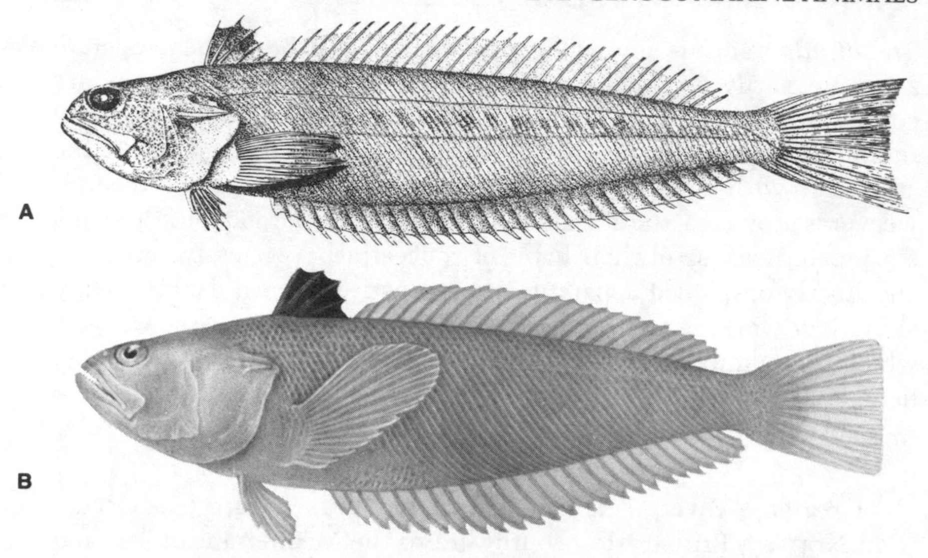

Figure 65. A. **Greater Weever,** *Trachinus draco* Linnaeus (From Joubin). B. **Lesser Weever,** *Trachinus vipera* Cuvier.

consciousness. In most instances, morphine fails to give relief. Untreated, the pain commonly subsides within 24 hours. Tingling, followed by numbness, develops about the wound. The skin about the wound at first is blanched, but soon becomes reddened, hot, and swollen. The swelling may be quite extensive and continue for ten days or longer. Other symptoms consist of headache, fever, chills, delirium, nausea, vomiting, dizziness, sweating, cyanosis, joint aches, loss of speech, slow heart beat, palpitation, mental depression, convulsions, difficulty in breathing, and death. Secondary infections are common in cases improperly treated. Gangrene has been known to develop as a complication. Recovery may take from several days to several months, depending upon the amount of venom received, condition of the patient, and other factors.

Figure 66. A. Head of *Trachinus*, showing the dorsal and opercular stings. B. Dorsal spine of *Trachinus vipera*, showing grooves in which the venom glands are located. Side and posterior views. C. Operculum of *Trachinus vipera* showing the daggerlike spine to which the venom gland is attached. Side view and cross section. D. Lateral view of left opercular sting of *Trachinus draco*. The integumentary sheath has been removed to show the venom glands which are seen as triangular masses lying above and below the shaft of the opercular spine. ▶

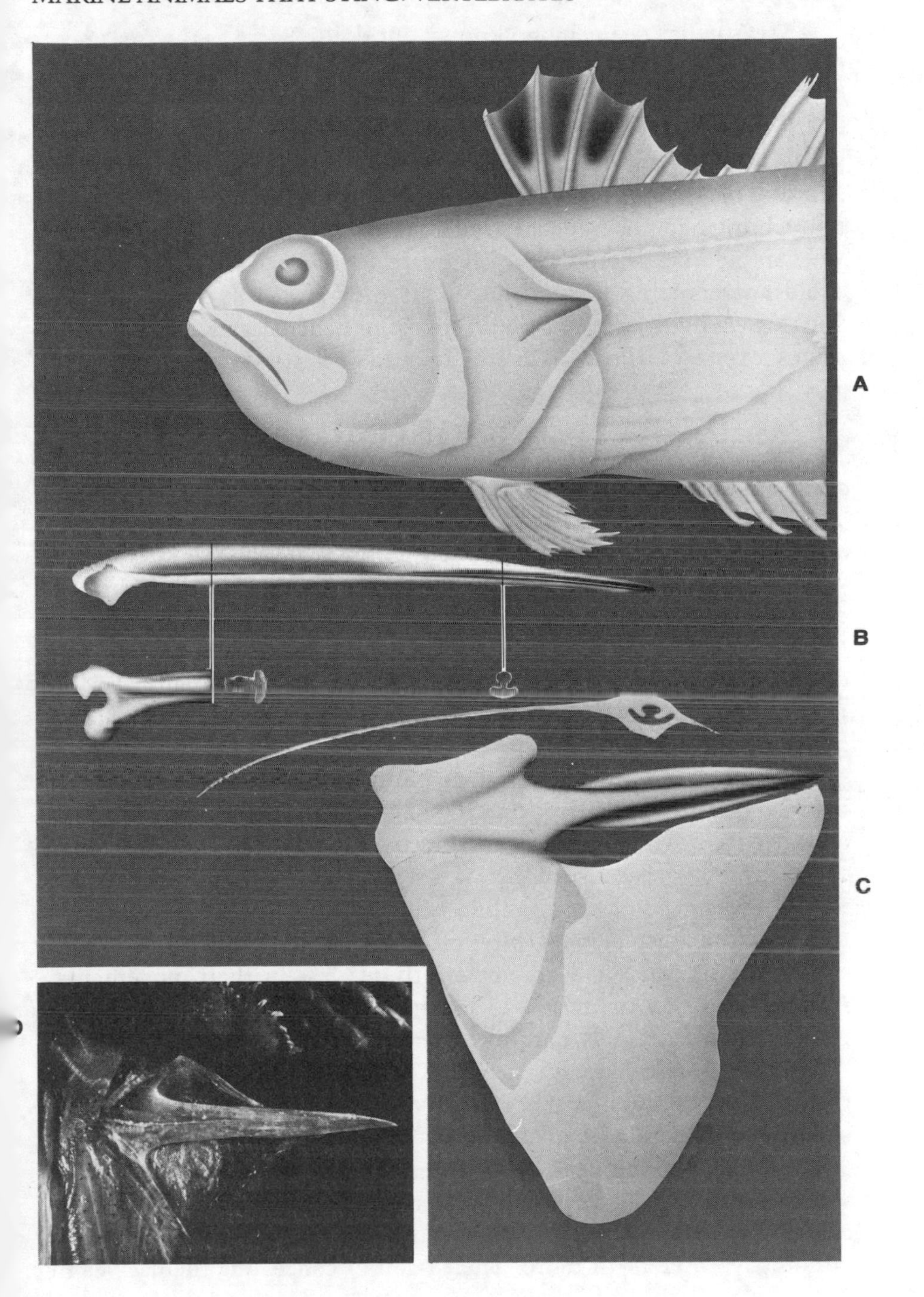
A
B
C

Treatment: There are no known antidotes. Same as the treatment of fish stings.

Prevention: Weeverfish stings are most commonly encountered while wading or swimming along sandy coastal areas of the eastern Atlantic or Mediterranean seas. Weevers are usually encountered partially buried in the sand or mud. Persons wading in waters where weevers abound should wear adequate footwear. Skin divers should avoid antagonizing these fishes since they are easily provoked into stinging. A living weever should never be handled under any circumstaces. Even when dead, weevers can inflict a nasty wound.

SCORPIONFISHES

Members of the family *Scorpaenidae,* the scorpionfishes, are widely distributed throughout all tropical and temperate seas. A few species are also found in arctic waters. Many scorpaenidae attain large size and are valuable food fishes, whereas others are relatively small and of no commercial value. Some species are extremely venomous.

Venomous scorpionfishes have been divided into three main groups on the basis of the structure of their venom organs, namely: (1) Zebrafish, (*Pterois*); (2) Scorpionfish proper (*Scorpaena*); and (3) the Stonefishes (*Synanceja*).

Zebrafish are among the most beautiful and ornate of coral reef fishes. They are generally found in shallow water, hovering about in a crevice or at times swimming unconcernedly in the open. They are also called turkeyfish because of their interesting habit of swimming around slowly and spreading their fanlike pectorals and lacy dorsal fins like a turkey gobbler displaying its plumes. These fish are frequently observed swimming in pairs, and are apparently fearless in their movements. Acceptance of an invitation to reach out and grab one of these fish results in an extremely painful experience because hidden under the "lace" are needlesharp fin stings. The fearlessness of the zebrafish makes it a particular menace to anyone working in its habitat, the shallow water coral reef areas.

Members of the genus *Scorpaena* are for the most part shallow water, bottom-dwellers, found in bays, along sandy beaches, rocky coastlines, or coral reefs, from the intertidal zone to depths of 50 fathoms (93 meters) or more. Their habit of concealing themselves in crevices, among debris, under rocks, or in seaweed, together with their protective coloration which blends them almost perfectly into their surrounding environment, makes them difficult to see. When they are removed from the water, they have the defensive habit of erecting their spinous dorsal fin and flaring out their armed gill covers, pectoral, pelvic and anal fins. The pectoral fins, although dangerous in appearance, are unarmed.

Stonefishes are largely shallow-water dwellers, commonly found in tidepools and shoal reef areas. *Synanceja* has the habit of lying motionless in coral crevices, under rocks, in holes, or buried in sand and mud. They appear to be fearless and completely disinterested in the careless intruder.

SPECIES OF SCORPIONFISHES

ZEBRAFISH TYPE

Zebrafish, Lionfish or **Turkeyfish,** *Pterois volitans* (Linnaeus) (Figure 67). Inhabits the Red Sea, Indian Ocean, China, Japan, Australia, Melanesia, Micronesia, and Polynesia. One of several closely allied species found around coral reefs.

SCORPIONFISH TYPE

Bullrout or **Sulky,** *Apistus carinatus* (Bloch and Schneider) (Figure 67). Inhabits the coast of India, Netherlands Indies, Philippine Islands, China, Japan, and Australia.

Waspfish or **Fortescue,** *Centropogon australis* (White) (Figure 68). Inhabits New South Wales and Queensland, Australia.

Bullrout, *Notesthes robusta* (Günther) (Figure 68). Inhabits New South Wales and Queensland, Australia.

Scorpionfish, *Scorpaena guttata* Girard (Figure 68). Ranges from central California south into the Gulf of California.

Scorpionfish, *Scorpaena plumieri* Bloch (Figure 69). Inhabits the Atlantic coast from Massachusetts to the West Indies and Brazil. One of several closely related species found in this general region.

Scorpionfish, Rascasse, Sea Pig, etc., *Scorpaena porcus* (Linnaeus) (Plate 7). Inhabits the Atlantic coast of Europe from the English Channel to the Canary Islands, French Morocco, Mediterranean and Black Seas.

Scorpionfish, *Scorpaenopsis diabolus* (Cuvier) (Figure 69). Inhabits the Netherlands Indies, Australia, Melanesia, and Polynesia.

Figure 67. A. **Zebrafish,** *Pterois volitans* (Linnaeus). (From Hiyama) B. **Bullrout** or **Sulky,** *Apistus carinatus* (Bloch and Schneider).

Figure 68. A. **Waspfish** or **Fortescue,** *Centropogon australis* (White). (From Whitley) B. ▶ **Bullrout,** *Notesthes robusta* (Günther). (After Bleeker) C. **Scorpionfish,** *Scorpaena guttata* Girard.

A
B
C

COMPARISON OF THE VENOM ORGANS

Structure	*Pterois*	*Scorpaena*	*Synanceja*
Fin spines	Elongated, slender	Moderately long, heavy	Short, stout
Integumentary sheath	Thin	Moderately thick	Very thick
Venom glands	Small-sized, well-developed	Moderate-sized, very well-developed	Very large, highly developed
Venom duct	Not evident	Not evident	Well-developed

Rockfish, *Sebastes* spp. (Figure 69). Many of the rockfish species found along the coastal areas of North American are now known to possess a venom apparatus and should be handled with care.

Lupo, *Inimicus japonicus* (Cuvier) (Figure 69). Inhabits the coastal areas of Japan.

STONEFISH TYPE

Stonefish, *Choridactylus multibarbis* Richardson (Figure 69). Found along coastal areas of India, China, Philippine Islands, and Polynesia.

Hime-Okoze, *Minous monodactylus* (Bloch and Schneider) (Figure 70). Inhabits the South Pacific Islands, China, and Japan.

Deadly Stonefish, *Synanceja horrida* (Linnaeus) (Figure 70). Inhabits India, East Indies, China, Philippine Islands, and Australia. The stonefish is an extremely dangerous species.

Venom Apparatus of Scorpionfishes: The venom organs of scorpionfishes vary markedly from one group to the next. A comparison of some of the more important differences appears in the chart above (see also Figure 71).

Figure 69. A. **Scorpionfish,** *Scorpaena plumieri,* Bloch. (From Everymann and Seale) B. **Scorpionfish,** *Scorpaenopsis diabolus* (Cuvier). (From Hiyama) C. **Lupo,** *Inimicus japonicus* (Cuvier). D. Stonefish type, *Choridactylus multibarbis* Richardson. (From Day) E. **Rockfish,** *Sebastes marinus* (Linnaeus). Length 40 centimeters. (From Daubin) ▸

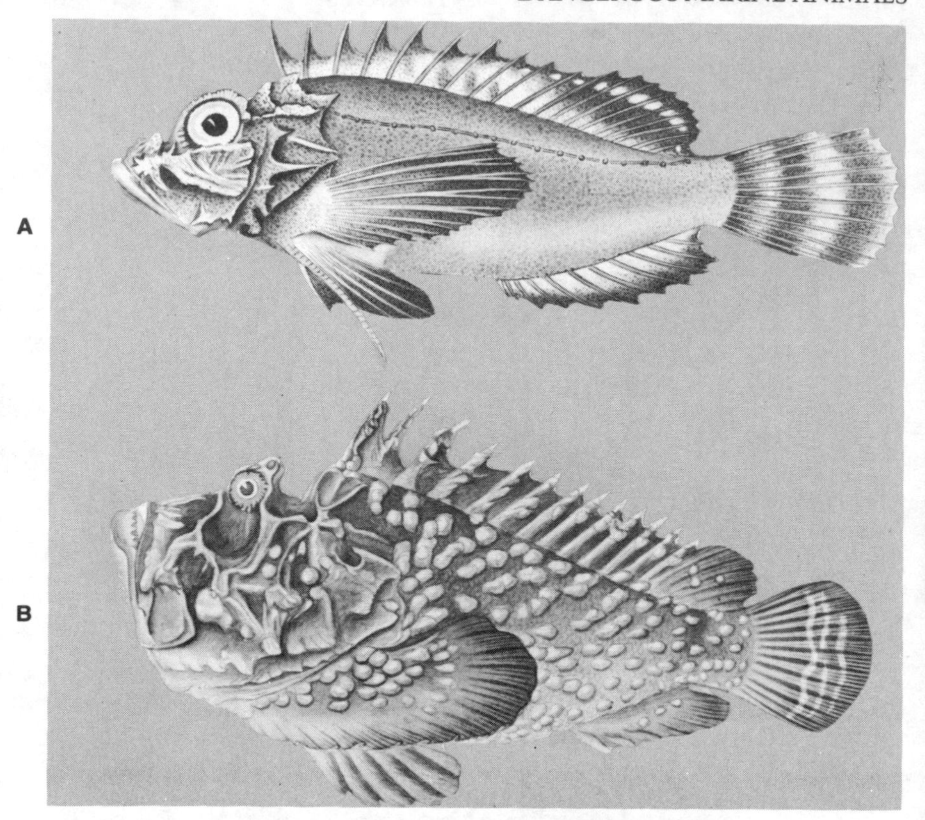

Figure 70. A. **Hime-okoze,** *Minous monodactylus* (Bloch and Schneider). B. **Deadly Stonefish,** *Synanceja horrida* (Linnaeus).

Zebrafish Type. The zebrafish, or *Pterois* type, has a venom apparatus consisting of 13 dorsal spines, 3 anal spines, 2 pelvic spines, and their associated venom glands. The spines are for the most part long, straight, slender, and camouflaged in delicate, lacy-appearing fins. Located on the front side of each spine are the glandular grooves, open on either side, which appear as deep channels extending the entire length of the shaft. Situated within these grooves are the venom glands. The glands are enveloped in a thin covering of skin—the integumentary sheath.

Scorpionfish Type. The scorpionfish, or *Scorpaena* type, have a variable number of dorsal spines, frequently 12, 3 anal and 2 pelvic spines, and their associated venom glands. The spines are shorter and heavier than those found in *Pterois*. The glandular grooves are restricted

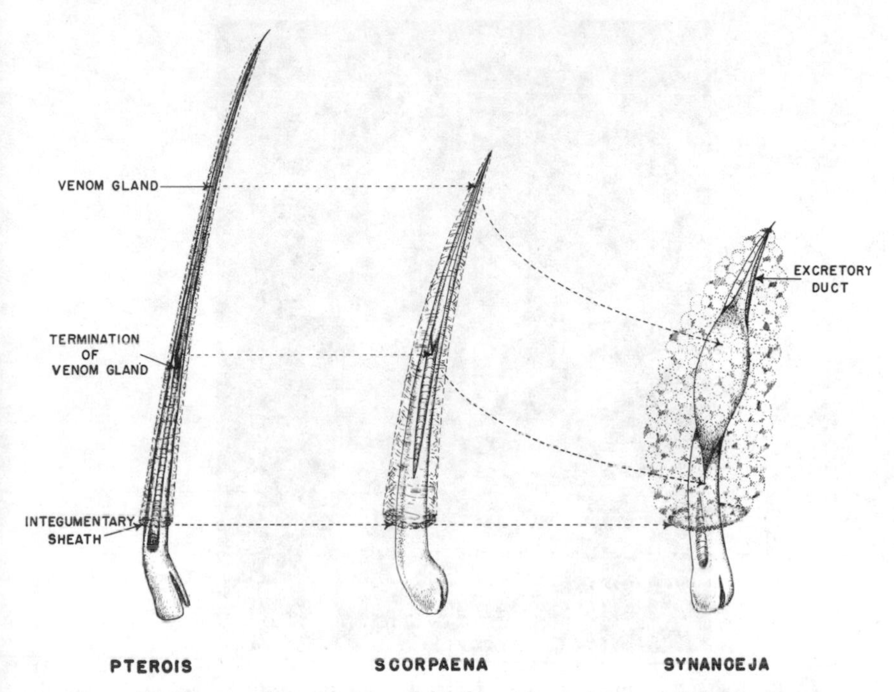

Figure 71. Showing a comparison of the three types of scorpionfish dorsal stings.

to about the distal two-thirds of the spine. The venom glands lie along the glandular grooves, but are limited to about the distal half of the spine. The enveloping integumentary sheath is moderately thick.

Stonefish type. The stonefish, or *Synanceja* type, usually have 13 dorsal spines, 3 anal spines, 2 pelvic spines, and their associated venom glands. The venom organs of this fish differ from the others by the short, heavy spines and greatly enlarged venom glands which are covered by a very thick layer of warty skin. This fish is particularly dangerous because of its camouflaged appearance—frequently resembling a large clump of mud or debris. The stings of *Synanceja* are difficult to detect because they are almost completely hidden (Figure 72).

Medical Aspects: Symptoms produced by the various species of scorpionfishes are essentially the same, varying in degree, rather than in quality. The pain is usually described as immediate, intense, sharp, shooting or throbbing, and radiates from the affected part. The area about the wound becomes ischemic, and then cyanotic. The pain produced by most scorpionfishes generally continues for only a few hours,

Figure 72. Photograph of anterior view of first dorsal sting of the **Stonefish,** *Synanceja horrida.* The integumentary sheath has been peeled back to show the large venom glands situated on either side of the spine shaft. Length of spine 1.3 inches (36 millimeters).

but wounds produced by *Synanceja* may be extremely painful and may continue for a number of days. Pain caused by *Synanceja* is sometimes so severe as to cause the victim to thrash about wildly, to scream, and finally, to lose consciousness. The area in the immediate vicinity of the wound gradually becomes cyanotic surrounded by a zone of redness, swelling, and heat (Plate 8). Subsequent sloughing of the tissues about the wound site may occur. In the case of *Synanceja* stings, the wound becomes numb, and the skin, some distance from the site of injury, becomes painful to touch. In some instances, complete paralysis of the limb may ensue. Swelling of the entire member that is affected may take place, frequently to such an extent that movement of the part is impaired. Other symptoms which may be present are: cardiac failure, delirium, convulsions, various nervous disturbances, nausea, vomiting, lymphangitis, swelling of the lymph nodes, joint aches, fever, respiratory distress, convulsions, and death. Complete recovery from a severe *Synanceja* sting may require many months, and may have an adverse effect on the general health of the victim.

Treatment: Same as treatment of venomous fish stings.

Prevention: Zebrafish stings are usually contracted by individuals who are attracted by the slow movements and lacy-appearing fins and who attempt to pick the fish up with their hands. Most scorpionfish stings result when an individual removes the fish from a hook, or nets, and is jabbed by their venomous spines. Stonefish are especially dangerous because of the difficulty of detecting them from their surroundings. Placing one's hands in crevices, or in holes inhabited by these fishes, should be done with caution. Knowledge of the habits and appearance of these fishes is most important.

TOADFISHES

The toadfishes are all members of the family *Batrachoididae*. All are small, bottom fishes which inhabit the warmer waters of the coasts of America, Europe, Africa, and India. With broad, depressed heads and large mouths, toadfishes are somewhat repulsive in appearance. Most of them are marine, but some are estuarine, or entirely freshwater, ascending rivers for great distances. They hide in crevices, burrows, under rocks, debris, among seaweed, or lie almost completely buried under a few centimeters of sand or mud. Toadfishes tend to migrate to deeper water during the winter months where they remain in a torpid state. They are experts at camouflage. Their ability to change to lighter or darker shades of color at will, and their mottled pattern make these fishes difficult to see.

SPECIES OF TOADFISHES

Toadfish, *Barchatus cirrhosus* (Klunzinger) (Figure 73). Inhabits the Red Sea.

Toadfish, or **Munda,** *Batrachoides grunniens* (Linnaeus) (Figure 73). Inhabits the coasts of Ceylon, India, Burma, and Malaya.

Toadfish, *Batrachoides didactylus* (Bloch) (Figure 73). Inhabits the Mediterranean Sea and nearby Atlantic Coasts.

Toadfish, Oysterfish, *Opsanus tau* (Linnaeus) (Figure 74). Found along the Atlantic Coast of United States, from Massachusetts to the West Indies.

Toadfish, Sapo, *Thalassophryne maculosa* Günther (Figure 74). West Indies.

Figure 73. A. **Toadfish,** *Barchatus cirrhosus* (Klunzinger). B. **Toadfish,** *Batrachoides grunniens* (Linnaeus). (From Sauvage) C. *Batrachoides didactylus* (Bloch). (From Steindachner)

Toadfish, Bagre Sapo, Sapo, *Thalassophryne reticulata* Günther (Figure 74). Inhabits the Pacific Coast of Central America.

In addition to the above, there are several other closely related species inhabiting certain coastal areas of Central and South America.

Venom Apparatus: The venom apparatus of toadfishes consists of two dorsal fin spines, two gill cover spines, and their associated venom glands. The dorsal spines are slender and hollow, slightly curved, and terminate in sharp, needlelike points. At the base and tip of each spine is an opening through which the venom passes. The base of each dorsal spine is surrounded by a glandular mass from which the venom is produced. Each gland empties into the base of its respective spine (Figure 75). The operculum is also highly specialized as a defensive organ for the introduction of venom. The horizontal limb of the operculum is a slender, hollow bone which curves slightly and terminates in a sharp tip. Openings are present at each end of the spine for the passage of venom. With the exception of the outer tip, the entire gill spine is encased within a glistening, whitish, pear-shaped mass. The broad,

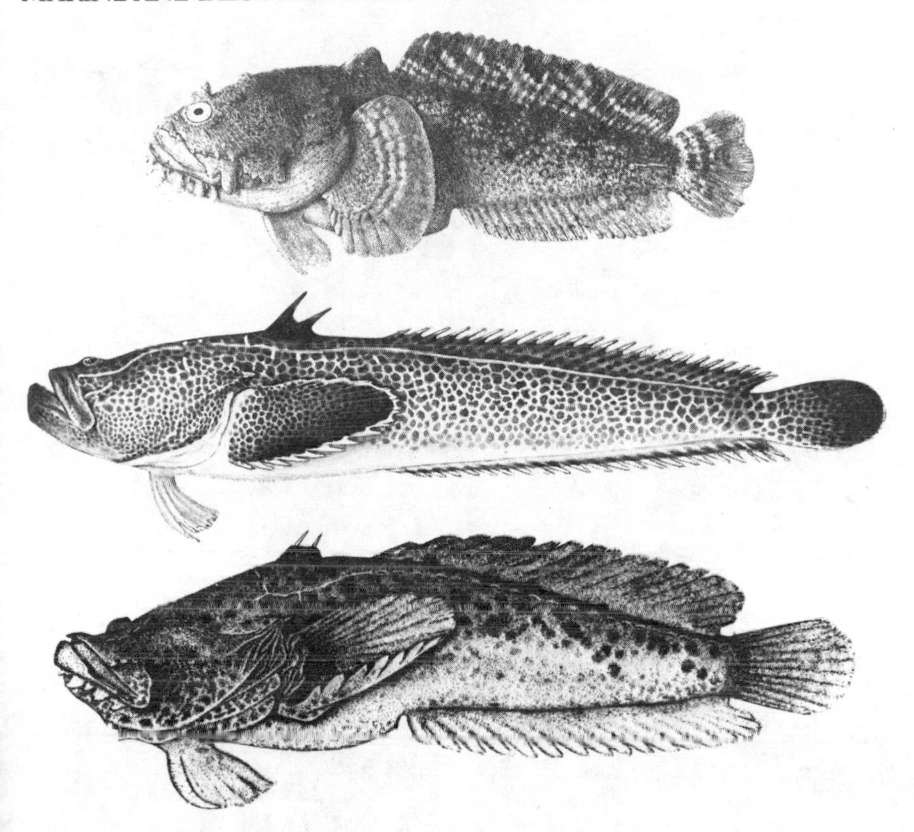

Figure 74. A. **Toadfish, Oysterfish,** *Opsanus tau* (Linnaeus). B. **Toadfish,** *Thalassophryne reticulata,* Günther. C. **Toadfish,** *Thalassophryne maculosa,* Günther. (From Collette)

rounded portion of this mass is situated at the base of the spine, and tapers rapidly as the tip of the spine is approached. This mass is the venom gland. The gland empties into the base of the hollow gill spine which serves as a duct.

Medical Aspects: The pain from toadfish wounds develops rapidly and is radiating and intense. Some have described the pain as being similar to that of a scorpion sting. The pain is soon followed by swelling, redness, and heat. No fatalities have been recorded in the literature but little else is known about the effects of toadfish venom.

Treatment: Same as treatment of stings from venomous fishes.

Prevention: To avoid stepping on toadfish, persons wading in waters inhabited by them should take the precaution to shuffle their feet

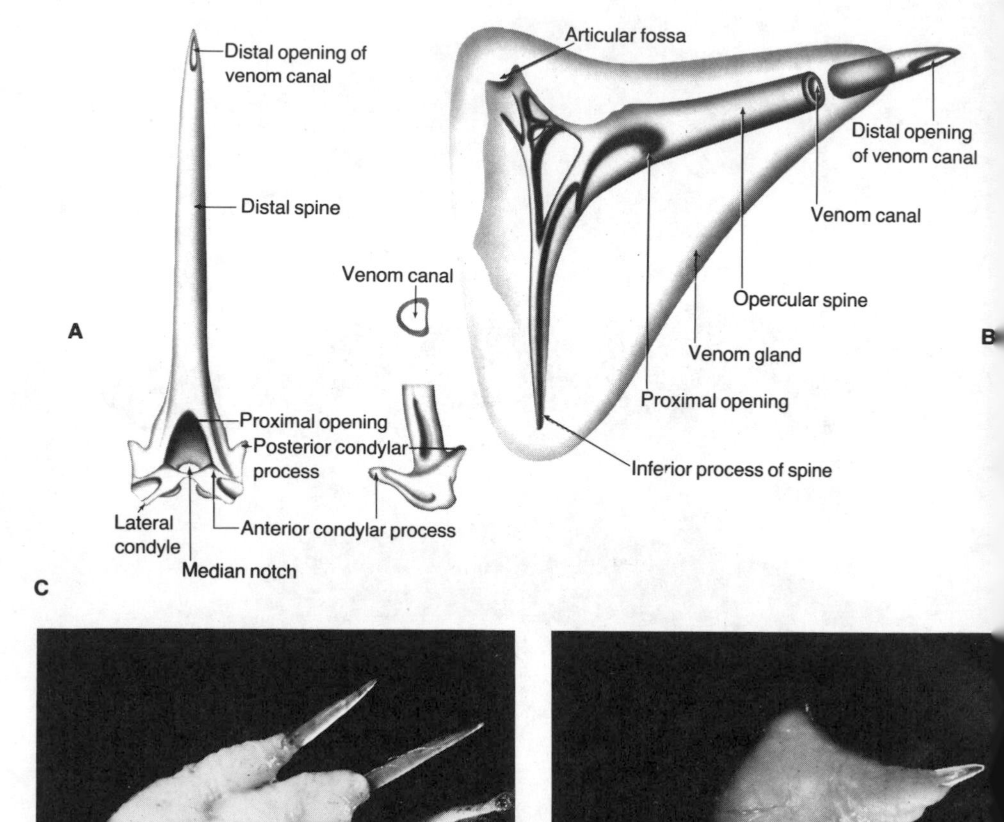

Figure 75. Anatomy of the venom apparatus of the **Toadfish,** *Thalassophryne dowi.* A. Opercular sting. B. Dorsal sting. C. Photograph of the dorsal stings. The integumentary sheath has been removed to expose the large venom glands at the base of the two dorsal spines. Length of spines 5/16 inch (11 millimeters). D. Left opercular sting. The skin has been removed to expose the venom gland. The opening of the venom canal is barely visible at the tip of the spine. Length of spine 3/8 inch (10 millimeters).

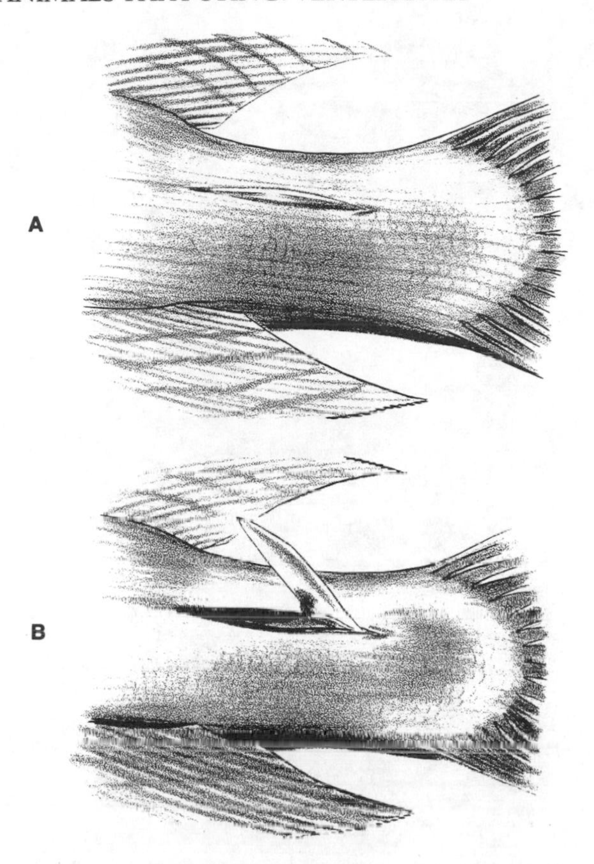

Figure 76. A. Tail of the surgeonfish, *Acanthurus*, in the contracted position. B. Caudal spine in the extended position. The fish can inflict a severe cut when the spine is extended.

through the mud. Removal of toadfishes from a hook or from nets should be done with care.

SURGEONFISHES

Surgeonfishes are members of the family *Acanthuridae*, and are reef-dwellers of warm seas. The principal genus of this family, *Acanthurus*, is characterized by the presence of a sharp, lancelike, movable spine on the side, at the base of the tail fin (Figure 76). When the fish becomes excited the spine, the point of which is directed forward, can be extended at right angles from the body of the fish. With a quick, lashing movement of the tail and the extended spine, large surgeonfishes are

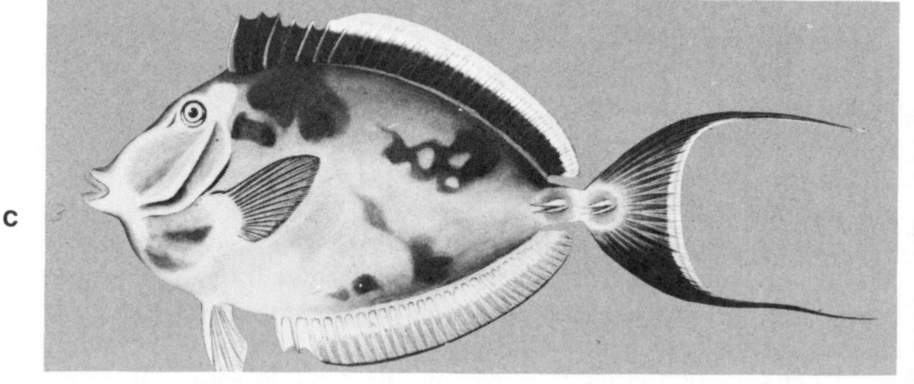

Figure 77. A. **Surgeonfish,** *Acanthurus xanthopterus,* Cuvier and Valenciennes. B. **Surgeonfish,** *Acanthurus bleekeri* Günther. C. **Surgeonfish,** *Naso lituratus* Bloch and Schneider. Three representatives species of surgeonfishes. Note the sharp spines near the base of the tail. In the genus *Acanthurus,* this spine can be extended, but *Naso* has spines which arise from plates and these are immovable. Both types can inflict serious wounds.

likely to inflict a deep and painful wound. Under normal circumstances, the spine remains adducted within a deep recess partially surrounded by an integumentary sheath. It has not been fully determined whether the spines of surgeonfishes are venomous. Some species appear to have venomous spines whereas others do not. Therefore, all surgeonfishes should be handled with care.

SPECIES OF SURGEONFISHES

Yellow-Finned Surgeonfish, *Acanthurus xanthopterus* Cuvier and Valenciennes (Figure 77). Indo-Pacific area.

Bleeker's Surgeonfish, *Acanthurus bleekeri* Günther (Figure 77). Found throughout the Indo-Pacific area, with the exception of Hawaii.

Surgeonfish, *Naso lituratus* Bloch and Schneider (Figure 77). Ranges from Polynesia to east Africa and the Red Sea.

Medical Aspects: Surgeonfish wounds are usually of the laceration type. Larger specimens are capable of inflicting deep and painful wounds. Secondary infections may occur.

Treatment: Same as for treatment of fish stings.

DRAGONETS

Dragonets are small, scaleless fishes with flat heads, bearing a preopercle that is armed with a strong spine. These fishes are brightly variegated in color, and have high and often filamentous dorsal fins. Some species are found in deep water, whereas others are shore fishes inhabiting shallow bays and reefs. The gill spine of the European species, *Callionymus lyra* Linnaeus (Figure 78), which is distributed along the Atlantic coast of Europe and the Mediterranean Sea, is said to be venomous. Dragonets are also considered to be capable of inflicting serious wounds. They should be handled with care.

RABBITFISHES

Rabbitfishes, which belong to the family *Siganidae,* are a group of spiny-rayed fishes which closely resemble the surgeonfishes. They differ from all other fishes in that the first and last rays of the pelvic fins are spinous. Usually valued as food, rabbitfishes are of moderate size, and abound about rocks and reefs from the Red Sea to Polynesia.

SPECIES OF RABBITFISHES

Rabbitfish, *Siganus fuscescens* (Houttuyn) (Figure 79). Found throughout the Indo-Pacific region.

Rabbitfish, *Siganus lineatus* (Valenciennes) (Figure 79). Inhabits the Philippines, Santa Cruz Islands, New Guinea, Solomon Islands, Australia, Okinawa, and the Ryukyu Islands.

Rabbitfish, *Siganus puellus* (Schlegel) (Figure 79). Inhabits the East Indies, Philippines, Palau, Gilbert, Marshall, and Solomon Islands.

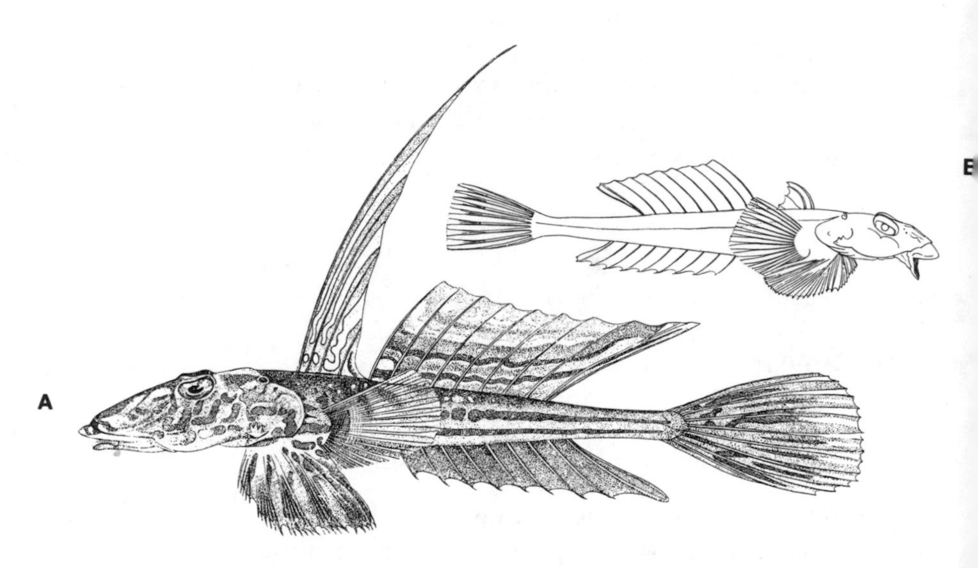

Figure 78. **Dragonet,** *Callionymus lyra* Linnaeus. A. Male. B. Female. Note preopercular spines which are believed to be venomous.

Venom Apparatus: The venom apparatus of *Siganus* consists of 13 dorsal, 4 pelvic and 7 anal spines, and their associated venom glands. A groove extends along both sides of the midline of the spine for almost its entire length. These grooves are generally deep and contain the venom glands which are located in the outer one-third of the spine, near the tip (Figure 80).

Figure 79. A. **Rabbitfish,** *Siganus fuscescens* (Houttuyn). B. **Rabbitfish,** *Siganus lineatus* (Valenciennes). C. **Rabbitfish,** *Siganus puellus* (Schlegel). ▸

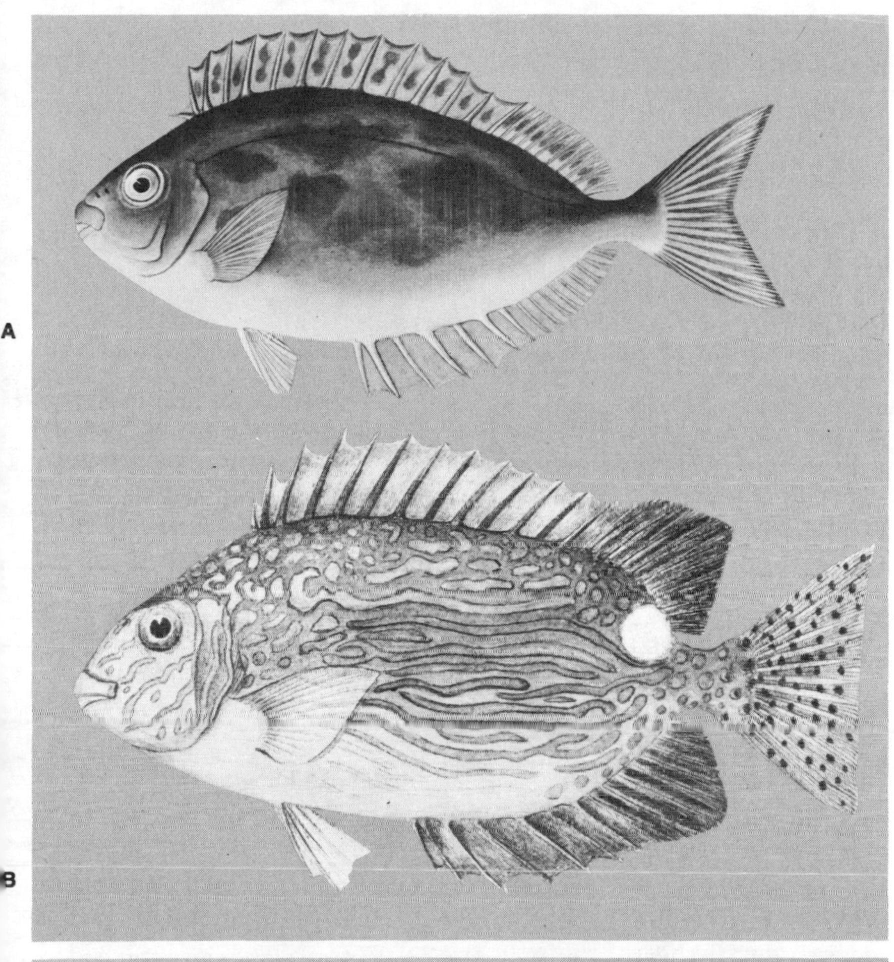
A
B

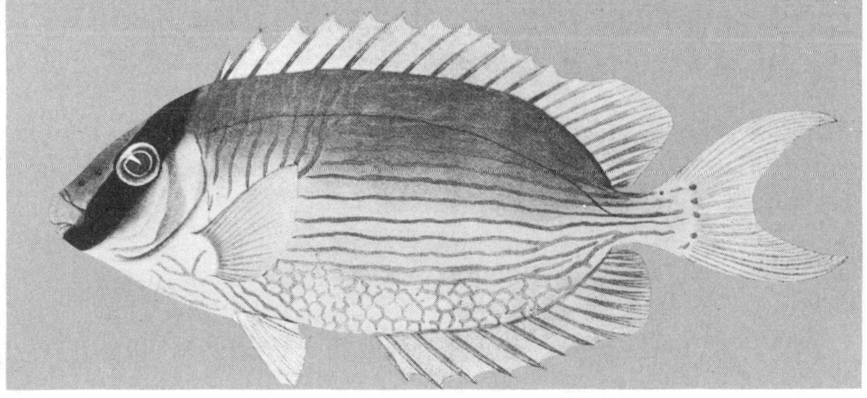

Medical Aspects: The symptoms produced by rabbitfish stings are said to be similar to those produced by scorpionfishes. There is no information available regarding the nature of the venom.

Treatment: Same as treatment of venomous fish stings.

Prevention: In handling rabbitfishes, one should be careful to avoid being jabbed with the dorsal, pelvic or anal stings.

STAR-GAZERS

Star-gazers are bottom-dwelling marine fishes, members of the family *Uranoscopidae,* having a cuboid head, an almost vertical mouth with fringed lips, and eyes on the flat upper surface of the head. Uranoscopids spend a large part of their time buried in the mud or sand with only their eyes and a portion of the mouth protruding.

SPECIES OF STAR-GAZERS

Star-Gazer or **Mishmimaokoze,** *Uranoscopus japonicus* Houttuyn (Figure 81). Inhabits southern Japan, southern Korea, China, Philippine Islands, and Singapore.

Star-Gazer, *Uranoscopus scaber* (Linnaeus) (Figure 81). Inhabits the eastern Atlantic and Mediterraean Sea.

Venom Apparatus: The venom apparatus of *Uranoscopus* is said to consist of two shoulder spines, one on either side, each of which protrudes through a sheath of skin (Figure 82). Venom glands are attached to these spines. The spine is said to have a double groove through which the venom flows.

Medical Aspects: There is no information available regarding the clinical characteristics of wounds produced by star-gazers. Wounds from the Mediterranean species, *Uranoscopus scaber,* may be fatal.

Treatment: Same as for treatment of other fish stings.

Prevention: Star-gazers should be handled with extreme care in order to avoid being jabbed by the shoulder spines.

LEATHERBACK

The leatherback, or lae, is a member of the family *Carangidae,* which includes the jacks, scad, and pompanos. A particular characteristic of this family of fishes is the presence of two separate spines in front of the anal fins. Several other species of carangids are believed to possess venomous spines, but the venom apparatus has been described in only the single species *Scomberoides sanctipetri* (Cuvier) (Figure 83). The leatherback is found throughout the tropical Indo-Pacific region. The carangids are a group of fast-swimming oceanic fishes which are generally encountered in the vicinity of coral reefs and islands.

Venom Apparatus: The venom apparatus consists of seven dorsal spines and two anal spines, their associated musculature, venom glands, and enveloping integumentary sheaths (Figure 84).

Medical Aspects: Stings from leatherback spines result in intense pain which may last for several hours. The anal spines are believed to inflict the most serious stings. The wounds are of the puncture-wound variety, and may be accompanied by redness and swelling.

Treatment: Same as treatment of fish stings. There are no specific antidotes available.

Prevention: Handle leatherbacks with care and avoid contact with the spines when removing them from a hook.

TREATMENT OF VENOMOUS FISH STINGS

Efforts in treating venomous fish stings should be directed toward achieving three objectives: (1) alleviating pain, (2) combating effects of the venom, and (3) preventing secondary infection. The pain results from the effects of the trauma produced by the fish spine, venom, and the introduction of slime and other irritating foreign substances into the wound. In the case of stingray and catfish stings, the retrorse barbs of the spine may produce severe lacerations with considerable trauma to the soft tissues. Wounds of this type should be promptly irrigated, or washed out with cold salt water or sterile saline if such is available. Fish stings of the puncture-wound variety are usually small in size, and

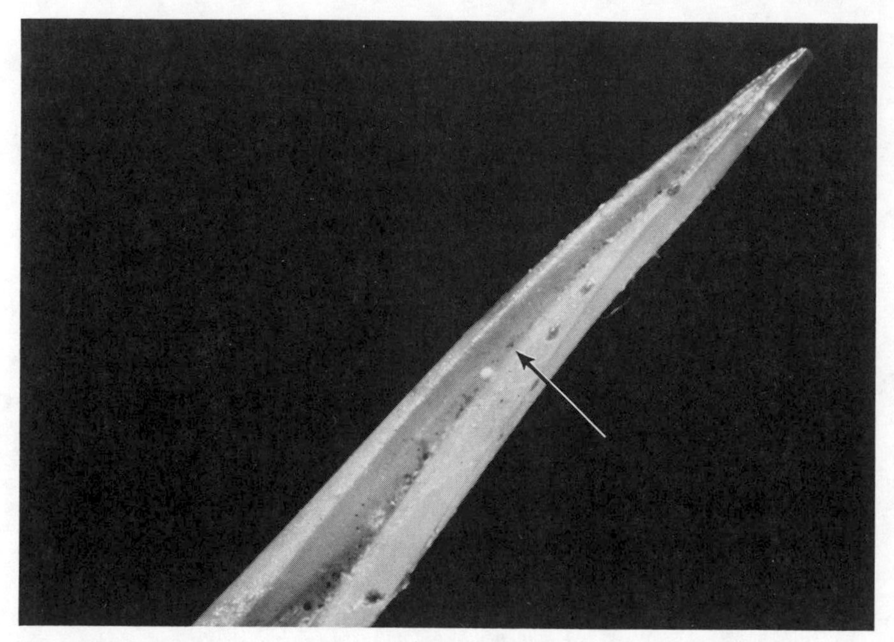

Figure 80. Venom apparatus of the **Rabbitfish,** *Siganus fuscescens.* Tip of the fifth dorsal sting of *Siganus fuscescens.* The skin has been removed to expose the venom gland (see arrow). The venom gland appears as a strip of tissue lying within the groove of the spine. Length of spine is 5/8 inch (15 millimeters).

Figure 81. A. **Star-gazer,** *Uranoscopus japonicus* Houttuyn. (From Temminck and Schlegel) B. **Star-gazer,** *Uranoscopus scaber* Linnaeus.

Figure 82. A. head of *Uranoscopus scaber.* Note the large spines protruding from the area just above the pectoral fin. Wounds from these spines can be fatal. B. Venom apparatus of the **Star-gazer,** *Uranoscopus scaber,* showing the left cleithral spine. Anatomical drawing. C. Actual fish. The skin has been removed to show the spine. The venom gland is barely visible at the base of the spine. ▸

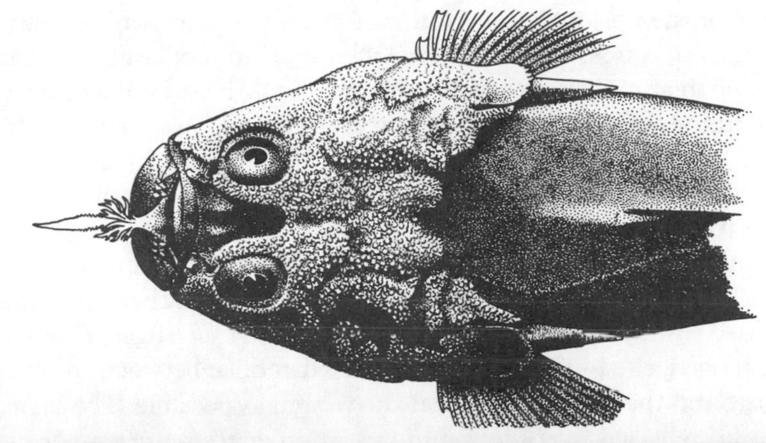

A

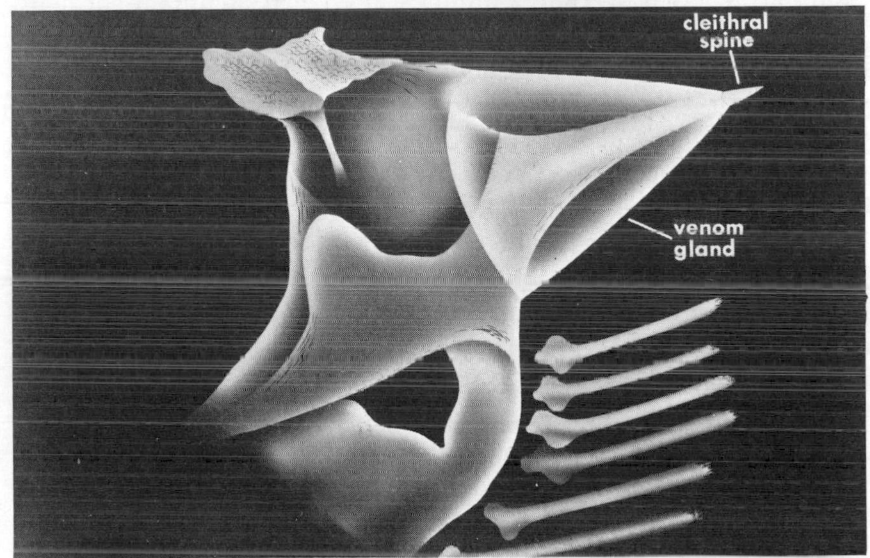

B

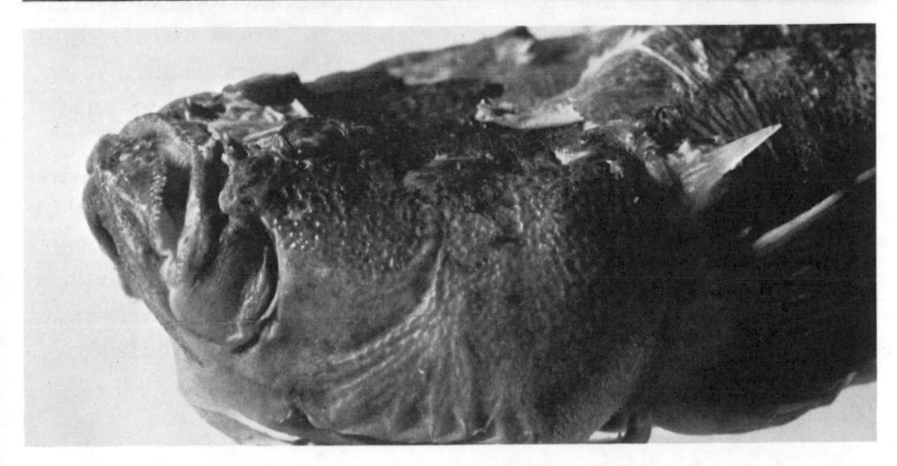

C

removal of the poison is more difficult. It may be necessary to make a small incision across the wound, and then apply immediate suction, and possibly irrigation. At any rate, the wound should be sucked promptly in order to remove as much of the venom as possible. However, it should be kept in mind that fishes do not inject their venom in the manner employed by venomous snakes, so at best, results from suction will not be too satisfactory.

There is a division of opinion as to the advisability and efficacy of using a ligature in the treatment of fish stings. If more than 10 minutes has lapsed since the sting was made, it is doubtful that a ligature is of any value. If used, the ligature should be placed at once between the site of the sting and the body, but as near the wound as possible. The ligature should be released every few minutes in order to maintain adequate circulation. Most doctors recommend soaking the injured member in hot water for 30 minutes to one hour. The water should be maintained at as high a temperature (120°F or 50°C) as the patient can tolerate without injury, and the treatment should be instituted as soon as possible. If the wound is on the face or body, hot moist compresses should be employed. The heat may have an attenuating effect on the venom since heating readily destroys stingray venom *in vitro*. Intravenous calcium gluconate injections are sometimes helpful. The addition of magnesium sulfate or epsom salts to the water is believed to be useful. Infiltration of the wound area with 0.5—2 percent procaine has been used with good results. If local measures fail to prove satisfactory, intramuscular or intravenous demerol will generally be efficacious. Following the soaking procedure, debridement and further cleansing of the wound may be desirable. Lacerated wounds should be closed with dermal sutures. If the wound is large, a small drain should be left in it for a day or two. The injured area should be covered with an antiseptic and sterile dressing.

Prompt institution of the recommended treatment usually eliminates the necessity of antibiotic therapy. If delay has resulted to any extent, the administration of antibiotics may be desirable. A course of tetanus antitoxin is an advisable precautionary measure.

The primary shock which follows immediately after the stinging generally responds to simple supportive measures. However, secondary shock resulting from the action of stingray venom on the cardiovascular system requires immediate and vigorous therapy. Treatment should be directed toward maintaining cardiovascual tone and the prevention of any further complications. Respiratory stimulants may also be required.

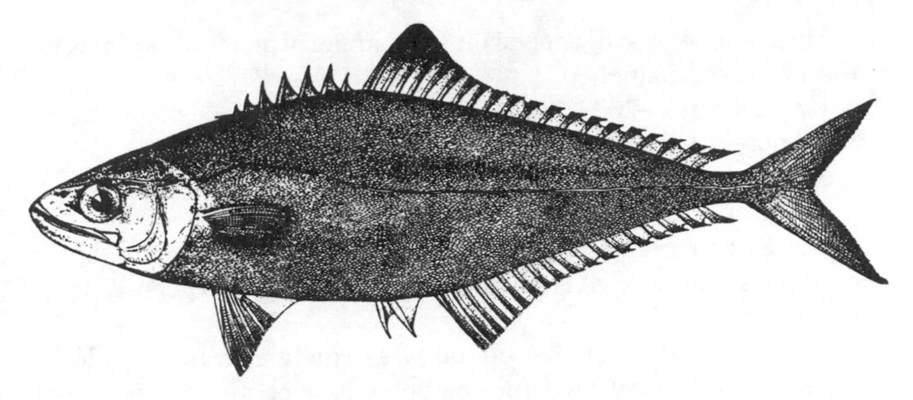

Figure 83. **Leatherback,** *Scomberoides sanctipetri* (Cuvier).

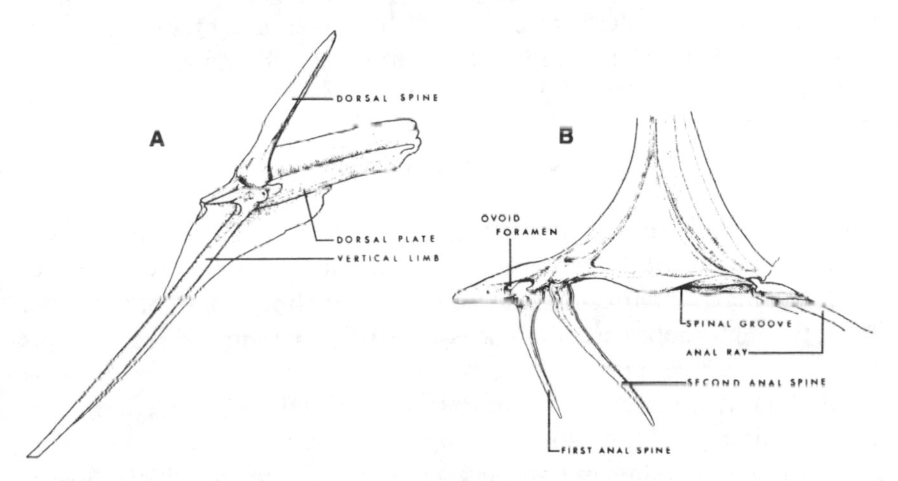

Figure 84. Venom apparatus of the **Leatherback,** *Scomberoides sanctipetri.* A. Dorsal spine. Length of spine 5/16 inch (8 millimeters). B. Anal spine. Length of second anal spine 1/2 inch (12 millimeters). The venom glands are embedded within the integumentary (skin) sheath surrounding the spines and are not visible to the naked eye.

Stonefish (*Synanceja*) stings are extremely painful and can be lethal. Fortunately, the Commonwealth Serum Laboratories, Melbourne, Australia, have developed an effective antivenin. This antivenin should also be considered for the treatment of envenomations by other scorpionfish species. The antivenin is prepared by hyperimmunizing horses with the stonefish venom. One unit of stonefish antivenin neutralizes 0.01 milligram of stonefish venom (1000 units will neutralize 10 milligrams of venom). Each stonefish spine produces about 5-10 milligrams of venom.

The initial dose will depend on the number of puncture wounds the patient received, namely,

1-2 punctures—2000 units (1 ampule)
3-4 punctures—4000 units (2 ampules)
5-6 punctures—6000 units (3 ampules)
and so on.

The antivenin should be given by intramuscular injection, but in severe cases should be given intravenously. If symptoms persist, repeat the initial dosage.

Local infiltration of the wound with emetine hydrochloride (65 milligrams/milliliters) procaine, or potassium permanganate (5 percent), may be helpful in relieving the pain, but generally soaking the limb in hot water is effective. Stonefish antivenin is produced in ampules of 2000 units (approximately 2 milliliters). The antivenin should be stored, protected from light, at 40-60°F or 2-8°C.

SEA SNAKES

Sea snakes are aquatic inhabitants of the tropical Pacific and Indian Oceans, ranging for the most part from the Samoan Islands westward to the East Coast of Africa, Japan to the Persian Gulf, along the coasts of Asia, through Indo-Australian seas to Australia. One species, *Pelamis platurus,* has an enormous geographical range extending from the west coast of Latin American across the Pacific and Indian Oceans to the east coast of Africa, and from southern Siberia to Tasmania. With the exception of a single freshwater species, *Hydrophis semperi,* which lives in freshwater Lake Bombon (also called Lake Taal), Luzon, Philippine Islands, all are marine. *Laticauda crockeri* is limited to the brackish water of Lake Tungano, Rennel Island, Solomon Islands. There are about fifty known species of sea snakes.

Sea snakes generally prefer sheltered coastal waters and are particularly fond of river mouths. Around shore, sea snakes may inhabit rock crevices, tree roots, coral boulders, or pilings. In regions where sea snakes are plentiful, more than 100 snakes may be taken by fishermen in a single net haul. A factor governing the distribution of most sea snakes seems to be the depth of water in which they feed. The depth must be shallow enough for them to go to the bottom to feed and to rise to the surface for air. However, sea snakes have been observed 100 to 150 miles (160 to 240 kilometers) from land. A large group of Stoke's sea snake (*Astrotia stokesi*) was once observed migrating between the Malay Peninsula and Sumatra. The mass of snakes was estimated to be about 10 feet (3 meters) wide and at least 60 miles (96 kilometers) long, consisting of several million individuals. Sea snakes are said to be the most abundant of all of the reptiles.

With their compressed, oarlike tails, sea snakes are well adapted for locomotion in the marine environment. Swimming is accomplished by lateral undulatory movements of the body. Sea snakes are able to float, lying motionless for long periods of time. They are also able to move backward or forward with amazing rapidity but when placed on land, are very awkward and move about with difficulty.

The bodies of sea snakes are covered with scales and have no limbs, ear openings, sternum, or urinary bladder. The bodies of sea snakes are more or less compressed posteriorly with a flat, paddle-shaped tail. Their eyes are immobile, covered by transparent scales, and are without lids. The tongue is slender, forked, and protrusile. Sea snakes are equipped with a well-developed venom apparatus.

Sea snakes are able to remain submerged for hours, but are unable to utilize oxygen from the water as fish do. They are equipped with an extended right lung, but the left lung is either vestigial or absent. Sea snakes capture their food underwater, usually swallowing the fish head first. They feed on or near the bottom, around rocks, holes, or crevices where they capture eels and other small fishes which are promptly killed with a vigorous bite of their venomous jaws.

The disposition of sea snakes is a subject of controversy, but of practical importance to those coming in contact with them. After reviewing the experiences of numerous writers and divers, one can only conclude that the docility of a sea snake varies with the species, the season of the year, and the manner in which the snake is approached. Generally speaking, sea snakes tend to be docile, but under some circumstances may be quite aggressive. One should not become over confident and should always be aware that a set of fully functional fangs and an extremely virulent venom (more toxic than king cobra venom) accompanies an apparent gentle nature. Bites are usually contracted while handling nets, sorting fish, wading, washing, or accidentally stepping on the snakes. Most species of sea snakes are 3-4 feet long (.9-1.2 meters) but may attain a length of 9 feet (2.7 meters) or more.

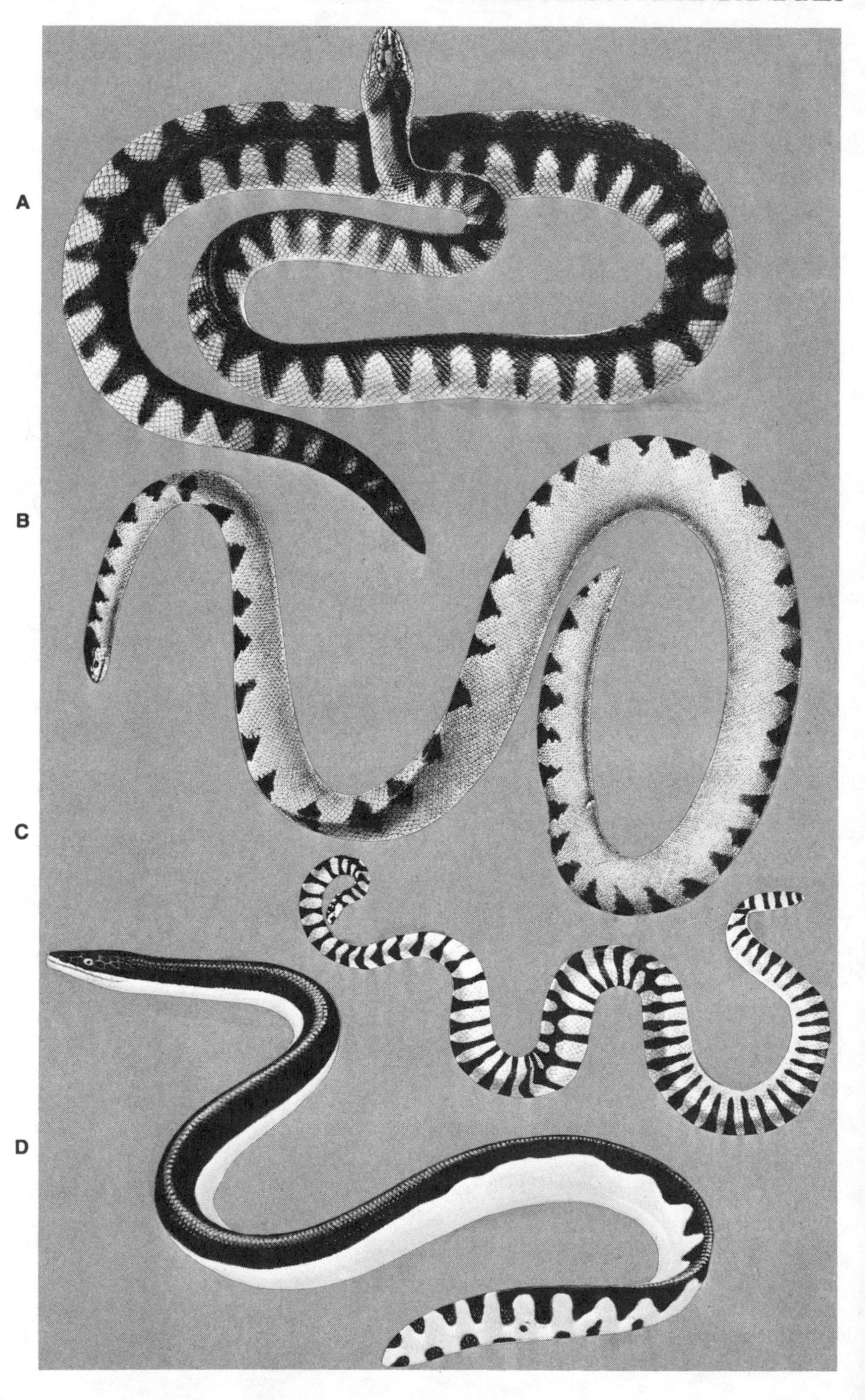
A
B
C
D

SPECIES OF SEA SNAKES

There are about fifty different species of sea snakes. The following species are reputed to be particularly dangerous to man.

Sea Snake, *Enhydrina schistosa* (Daudin) (Figure 85). Inhabits the Persian Gulf to Cochin China and north coast of Australia.

Chittul, or **Banded Sea Snake,** *Hydrophis caerulescens* (Shaw) (Figure 85). Ranges from the Persian Gulf to Japan, and the Netherlands Indies.

Sea Snake, *Hydrophis nigrocinctus* (Daudin) (Figure 85). Inhabits the Bay of Bengal.

Hardwick's Sea Snake, *Lapemis hardwicki* (Gray) (Figure 86). Ranges from southern Japan to the Merguri Archipelago, and along the coast of northern Australia.

Yellow-Bellied Sea Snake, *Pelamis platurus* (Linnaeus) (Figure 85). This species has the greatest range of any sea snake, extending from East Africa, throughout the Indo-Australian area, eastward to the Gulf of Panama.

Venom Apparatus: Sea snakes inflict their wounds with the use of fangs, which are reduced in size, but are of the cobra type (Plate 8). In comparison with other venomous snakes, sea snake dentition is relatively feeble but, nevertheless, fully developed for venom conduction. The venom apparatus consists of the venom glands and fangs. The venom glands are situated, one on either side, behind and below the eye, and in front of the tympanic bones. They are ovate and have an elongated venom duct which terminates at the base of the fangs. Most sea snakes have two fangs on each side but some have only one. The venom duct enters at the base of the fang through a relatively broad, triangular opening of the canal. Sea snake venom has neurotoxic, myotoxic, and hemotoxic properties. One drop (.03 milliliters) contains sufficient poison to kill three adult men. Some species can inject up to 8 drops in a single bite.

Medical Aspects: Symptoms caused by the bite of a sea snake characteristically develop rather slowly, taking from 20 minutes to several hours, but usually about one hour, before definite symptoms begin. Aside from the initial prick, there is no pain or reaction at the site of the bite. The victim may even fail to connect the bite with his illness.

◂ Figure 85. **Sea Snake,** *Enhydrina schistosa* (Daudin). B. **Chittul** or **Banded Sea Snake,** *Hydrophis caerulescens* (Shaw). C. **Sea Snake,** *Hydrophis nigrocinctus* (Daudin). D. **Yellow-bellied Sea Snake,** *Pelamis platurus* (Linnaeus).

In some instances, the initial generalized symptoms are a mild euphoria, whereas in others there is aching and anxiety. A sensation of thickening of the tongue and a generalized feeling of stiffness of the muscles gradually develops. Common complaints during the early stage are "aching," "stiffness," or pain upon movement. There may be little indication of actual weakness at this time. The paralysis which soon follows is usually generalized, but of the ascending type, beginning with the legs, and, within an hour or two, involving the trunk, arm and neck muscles. Lockjaw is one of the outstanding symptoms. Drooping of the eyelids is an early and characteristic sign. The pulse becomes weak,

Figure 86. **Hardwick's Sea Snake,** *Lapemis hardwicki* Gray.

irregular, and the pupils dilate. Speaking and swallowing become increasingly difficult. Thirst, burning, or dryness of the throat may also be present. Nausea and vomiting are not uncommon. Muscle twitchings, twisting movements, and spasms have been noted. Ocular and facial paralysis may later develop. In severe intoxication, the symptoms become progressively more intense, the skin of the patient is cold, clammy, cyanotic, convulsions begin and are frequent, respiratory distress becomes very pronounced, and finally the victim succumbs in an unconscious state. The overall case fatality rate has been estimated to be about 28 percent.

Treatment: If treatment is to be effective it must be instituted immediately following the bite. First aid and treatment should generally be directed toward:

(1) removing as much venom as possible from the wound;
(2) retarding absorption of the venom;
(3) neutralizing the venom;
(4) mitigating the effects produced by the venom; and
(5) preventing complications, including secondary infections.

First Aid: Absorption of sea snake venom is rapid. In most instances the venom is absorbed before first aid can be administered. Suction is of value only if it can be applied within the first few minutes following the bite. It is generally advisable to leave the bite alone. The affected limb should be immediately immobilized, and *all exertion must be avoided.* The patient should lie down and keep the immobilized part below the level of the heart. A tourniquet should be applied tight enough to occlude the superficial venous and lymphatic return. The tourniquet should be applied to the thigh in leg bites or above the elbow in upper limb bites. It should be released for 90 seconds every 10 minutes. A tourniquet is of little value if applied later than 30 minutes following the bite, and it should not be used for more than 4 hours. The tourniquet should be removed as soon as antivenin therapy has been started. Some workers believe that the tourniquet is of little or no value in sea snake bites. If sea snake antivenin, or a polyvalent antivenin containing a krait (*Elapidae*) fraction is available, it should be administered intramuscularly either in the buttocks or at some other side distant from the bite. The antivenin should be given only after the appropriate skin or conjunctival test has been made. There appears to be an increasing number of physicians who recommend injecting the antivenin in adequate quantities intravenously without concern for side reactions. They believe that any hypersensitivity can be controlled with the use of adrenalin. Usually one unit (vial or ampule) is sufficient until the patient can be transported to a physician. Keep the patient warm. The patient should not be given alcoholic beverages but may be given water, coffee, or tea. Keep the patient calm and reassured with encouraging words and actions. Transport the patient immediately to the nearest doctor or hospital, but do not allow him to walk or exert himself in any way. If at all possible place the offending dead snake in a container and give it to the physician for identification.

Therapy: Sea snake envenomation is a medical emergency requiring immediate attention. A delay in instituting proper medical treatment can lead to consequences far more tragic than might be incurred in an ordinary traumatic injury. The first step is to determine if envenomation has actually occurred. In most instances, by the time the patient reaches the physician, the 1-hour test period since the bite was made will have been exceeded so there should be some clinical evidence of poisoning if any is to develop. If the patient has been carefuly examined and there is no evidence of intoxication, the patient needs only reassurance and observation for a period of an hour or two. If after a period of observation there is no evidence of toxemia, the patient can be released.

If a period of more than one hour has elapsed and there is definite evidence of intoxication, antivenin therapy should be instituted immediately upon arrival at the hospital or physician's office. At this time, suction and incision are useless. There is now available a sea snake antivenin, which is prepared from the venom of *Enhydrina schistosa* and can be obtained from the Commonwealth Serum Laboratories, Melbourne, Australia, and the Snake and Venom Research Laboratories, Penang, Malaysia. This antivenin is concentrated and purified so as to minimize hypersensitivity reactions. If sea snake antivenin is not available, use a polyvalent antiserum containing a krait (*Elapidae*) fraction. A skin or conjunctival test should be done prior to injection of the antivenin. In persons with a history of extensive allergies, the antivenin should be injected with caution even in the presence of a negative skin test. Follow the instructions given in the brochure that accompanies the antivenin. If instructions are not available, these procedures can be used: The antivenin can be given either intramuscularly or intravenously. The intramuscular route is said to be safer, but less effective. Again, physicians are increasingly recommending injecting the antivenin intravenously without concern for hypersensitivity. A portion of the first ampule should be injected subcutaneously proximal to the bite, surrounding the wound, or in advance of the swelling. Antivenin should never be injected into a finger or toe nor should large amounts be injected into the injured part. A second portion of the antivenin should be injected intramuscularly into a large muscle mass at some distance from the bite, and the last portion should be given intravenously.

Intravenous antivenin is indicated with patients in shock. When given intravenously the antivenin should be given by the drip method over a period of 1 hour. The antivenin can be added to the physiological saline solution and given in a continuous drip. Subsequent doses can then be added to the saline solution.

The incidence of sensitivity reactions is said to be lessened when the antivenin is combined with hyaluronidase (10 milliliters of antivenin with 1,000 units of hyaluronidase) and given intramuscularly. It is recommended that an injection of 250 milligrams of cortisone be given intramuscularly prior to administering the antivenin, but not at the same time that the antivenin is given. If hypersensitivity reactions occur, they can usually be controlled with adrenalin subcutaneously with or without intravenous antihistaminic drugs. If the patient shows a positive reaction to the initial sensitivity test, the antivenin will have to be administered in graded doses in accordance with standard medical procedures for desensitization.

One ampule of sea snake antivenin neutralizes 10 milligrams of *Enhydrina schistosa* venom and the venom of other common sea snake species. The minimum effective dosage is 1 ampule. In severe poisoning as evidenced by ptosis, weakness of external eye muscles, dilation of pupils with sluggish light reaction, and leucocytosis exceeding 20,000, three to four or more ampules are required. Children generally respond satisfactorily to smaller doses of antivenin than that required by adults, contrary to previous textbook statements. If envenomation is present, antivenin therapy should be given as soon as possible, but it may be successful even 8 or more hours after the bite has occurred. Parenteral fluids should always be given following a severe envenomation. Cryotherapy is not recommended.

Supportive measures such as blood transfusions, plasma, vasopressor drugs, antibiotics, antitetanus agents, oxygen, etc., may be required. Corticosteroids are the drugs of choice in combating delayed allergic reactions provoked by the venom or the horse serum. Neurological disorders such as cerebral, meningeal, myelitic, radicular, and neuritic disorders (particularly of the fifth and sixth cervical roots) have been reported. Delayed reactions usually do not appear until 3 to 7 days after administration of the antivenin.

Sensitivity Tests: A sensitivity test must be carried out on all victims of sea snake poisoning before horse serum antivenin is administered. Directions for these tests will usually be found in the package containing the antivenin. In the absence of specific instructions, follow these steps:

1. Inject 0.10 milliliters of a 1:10 dilution of the horse serum or antivenin intracutaneously on the inner surface of the forearm. Use the specific hypodermic needle provided for the test. If one is not provided, use a short, 27-gauge needle. If the test is done correctly, a wheal will be raised at the site of the injection. the wheal is white at first but if the test is positive the area about the point of injection will become red within 10 to 15 minutes. If any local or systemic allergic manifestations develop within 20 minutes of the test, *do not* give antivenin. Leave this decision to the physician.

If the patient develops a severe reaction to the test (restlessness, flushing, sneezing, urticaria, swelling of the eyelids and lips, respiratory distress, or cyanosis), inject 0.3 to 0.5 milliliters of 1:1,000 adrenalin subcutaneously and observe the patient closely. Be prepared to administer artificial respiration. A cardiac stimulant may aso be needed if shock develops.

2. An alternative to the skin test is the eye test. One or two drops of a 1:10 solution of the horse serum or antivenin are placed on the conjunctiva of one eye. If the test is positive, redness of the conjunctiva will develop within a few minutes. If the reaction is very severe, it should be controlled by depositing a drop or two of 1:1,000 adrenalin directly on the conjunctiva.

3. If a serum sensitivity test is positive, desensitization should be carried out before administering antivenin.

Prevention: Although sea snakes are generally considered to be docile and at times reluctant to bite, some species are aggresive and have no biting inhibitions. It should be kept in mind that even the "harmless" snakes are fully equipped with a venom apparatus and a potentially lethal poison. Sea snakes thus merit respect and thoughtful consideration. Sea snakes may occasionally bite a bather. It is estimated that one sea snake bite occurs per 270,000 man-bathing-hours in an endemic area such an Penang, Malaysia. The most dangerous areas in which to swim are river mouths. It is probable that accidents are more likely to occur in a river mouth where the sea snakes are more numerous and the water more turbid. The turbidity of the water and resulting poor visibility for the sea snake may contribute to their becoming fearless and less discriminating in their biting habits. When possible, avoid swimming in a river mouth. When wading in an area inhabited by sea snakes, shuffle your feet. This will prevent stepping on them. The majority of sea snake bites have taken place among native fishermen removing fish from their nets. A person should be extremely careful in handling a net haul containing sea snakes. It is advisable first to remove the snakes with the use of a hooked stick or a wire before attempting to handle them. Sea snakes are occasionally captured while fishing with a hook and line. Do not attempt to remove the hook from the mouth of the snake. Cut the line, and let the snake drop into the water. Do not handle the snake.

V

MARINE ANIMALS POISONOUS TO EAT: INVERTEBRATES

Tropical waters attract the skin diver who is seeking new adventure. The waters are warm and clear, and the sealife is abundant. But venturing into strange tropical waters is not without some danger. In this chapter we are not concerned about what might harm us from without, but rather from the dangers within—those sea foods likely to be eaten by the diver while in the field. Fortunately, most sea foods are edible and nourishing. However, there are those known to contain some of the most hazardous food poisons known to science. These illnesses are not due to bacterial food poisoning, but to a group of marine animals whose flesh, under certain circumstances, may contain poisonous chemical substances.

CLASSIFICATION OF POISONOUS MARINE ANIMALS

There is a large number of poisonous marine animals known to marine biotoxicologists, but only a few general categories of them will be dealt with here. For the purposes of this book, marine animals that are poisonous to eat may be classified as follows:

POISONOUS INVERTEBRATES

A. **Molluscs** which have been feeding on toxic dinoflagellates can, if ingested, cause paralytic shellfish poisoning.

B. **Coelenterates**.

1. Sea anemones (*Rhodactis*).

2. Zoanthids (*Palythoa*).

C. **Echinoderms.**

1. Sea urchin eggs.
2. Sea cucumbers.

D. **Arthropods.**

1. Asiatic horseshoe crabs.
2. Crabs.

POISONOUS VERTEBRATES

A. **Fishes.** There are estimated to be more than 400 species of fishes which can produce an intoxication when ingested. Intoxications resulting from the ingestion of fish flesh is referred to as *ichthyosarcotoxism*. There are various types of *ichthyosarcotoxism*, but only a few of the more common poisonings will be presented in this manual. They include the following:

1. *Elasmobranch Poisoning*—due to the ingestion of poisonous sharks, rays, and some of their relatives.
2. *Ciguatera*—due to the ingestion of various species of tropical reef fishes. A related form of poisoning is known as *clupeitoxism* which is due to the ingestion of herringlike fishes of the Order *Clupeiformes*.
3. *Scombroid Poisoning*—due to the ingestion of inadequately preserved tunas and related species of scombroid fishes.
4. *Gempylid Poisoning*—due to the ingestion of the castor oil or gempylid fish.
5. *Hallucinogenic Fish Poisoning*—due to ingestion of certain tropical reef fishes which produce hallucinations.
6. *Puffer Poisoning*—due to the eating of the flesh of puffer fishes, members of the Order *Tetraodontiformes*.

B. **Reptiles.** Some of the most violent marine biotoxications known have been caused by the eating of the flesh of marine turtles of the genera *Chelonia, Eretmochelys,* and *Dermochelys.*

C. **Marine Mammals.** Ingestion of the flesh and various organs of certain marine mammals may cause violent intoxications. This group includes the polar bear, certain whales, dolphins, and porpoises.

Each of these groups of poisonous marine organisms will be discussed in greater detail in the two chapters that follow.

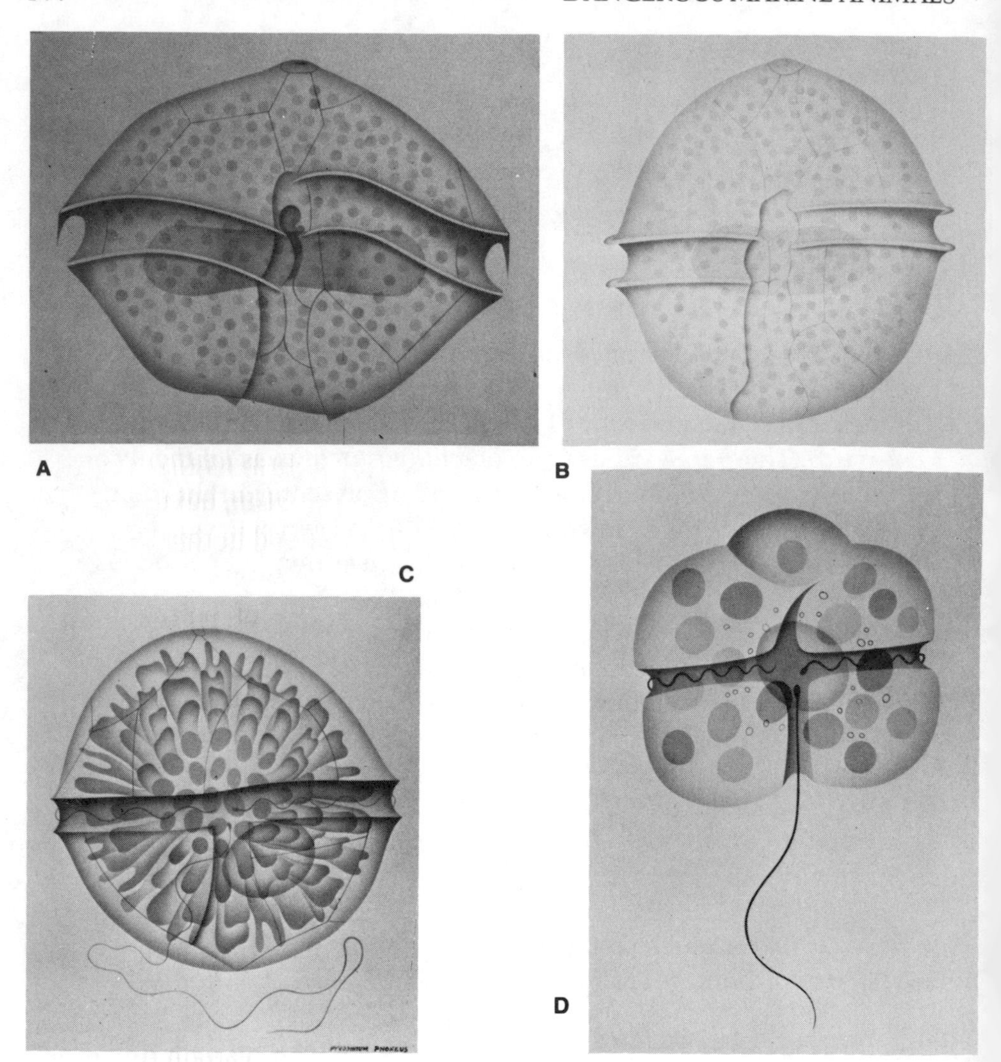

Figure 87. **Dinoflagellates** A. *Gonyaulax catenella* Whedon and Kofoid. X1440. B. *Gonyaulax tamarensis* Lebour. X1200. C. *Pyrodinium phoneus* Woloszynska and Conrad. X1980. These are the most commonly encountered dinoflagellates causing paralytic shellfish poisoning. *Pyrodinium phoneus* is now considered to be *Gonyaulax phoneus*. D. *Gymnodinum breve* Davis. X1440. This dinoflagellate may cause red tide and a mass mortality of fishes. It may also cause respiratory irritation in humans.

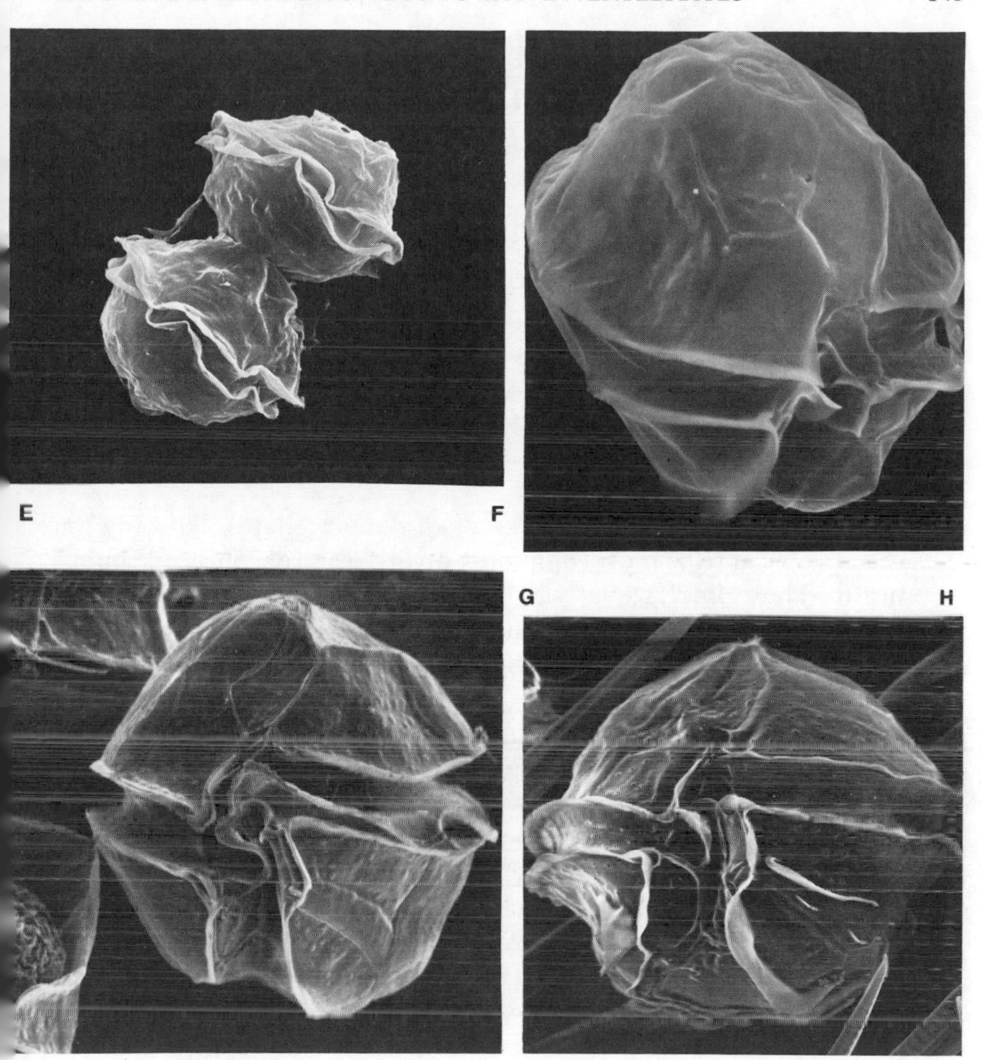

igure 87. **Dinoflagellates** *continued.* E. *Gonyaulax catenella* Whedon and Kofoid, isolated from a ed tide sample collected January, 1975, from Owase Bay, Mie Prefecture, Japan. Dorsal view of a wo-celled chain. X1440. F. *Gonyaulax tamarensis* Lebour, isolated from Lynher River, England, ollected June, 1957. Ventral view with cells right side more exposed. The toxicity of this species has een questioned. X2200. G. *Gonyaulax excavata* (Braarud), isolated from a red tide from Lanesville tation, Gloucester, Massachusetts. This is now believed to be the toxic causative agent of the New ngland red tide rather than *Gonyaulax tamarensis*. X2100. H. *Gonyaulax acatenella* Whedon and ofoid, isolated from a red tide sample from Malaspina Inlet, British Colombia, Canada. Taken in June, 1965. Reputed to be toxic. X1700. (Courtesy A.R. Loeblich, III, and L.A. Loeblich.)

POISONOUS MARINE INVERTEBRATES

PARALYTIC SHELLFISH POISONING

Paralytic shellfish poisoning occurs in human beings as a result of eating molluscs containing potent biotoxins derived from planktonic unicellular organisms known as dinoflagellates. These dinoflagellates are sometimes observed to "bloom" in vast numbers and may cause a phenomenon which is referred to as "red tide." At times red tide may cause a mass mortality of fishes. However, molluscs may become poisonous from feeding on dinoflagellates even when a red tide is not in evidence. Dinoflagellate blooms constitute a public health hazard because of their unpredictability and the rapidity with which toxic concentrations may develop. The precise cause of these blooms has not been fully determined.

There are many different kinds of dinoflagellates, but only four species appear to be most commonly involved and of concern to human health. These dinoflagellates bloom sporadically throughout the world, but seem to occur most frequently in north and south temperate latitudes. There are two species of dinoflagellates most commonly incriminated in human intoxications: *Gonyaulax catenella* Whedon and Kofoid (Figure 87), along the north Pacific coast of North America from central California, northward along the coasts of Oregon, Washington, British Columbia, Alaska, and westward along the Aleutian Islands to the coasts of Japan, and in the Southern Hemisphere along the coast of Chile; *Gonyaulax tamarensis*[1] Lebour (Figure 87), the most common toxic dinoflagellate along the northeast coast of North America including Massachusetts, New Hampshire, and Maine, and in Canada along the coasts of Nova Scotia, New Brunswick and Quebec, and across the Atlantic Ocean to the North Sea countries of Britain, West Germany, Belgium, and Denmark. *Pyrodinium phoneus* Woloszynska and Conrad[2] (Figure 87) has been incriminated in the Netherlands. Another dinoflagellate which is known to be toxic and may cause mass mortality among fishes, but may or may not be involved in human biotoxications, is *Gymnodinium breve* Davis (Figure 87). *Gamberdiscus toxicus* Adachi and Fukuyo has been incriminated as a causative agent of ciguatera fish poisoning. (See Chapter VI).

[1]According to some investigators *G. tamarensis* is actually non-toxic, and the correct identification is *G. excavata* Braarud.
[2]*P. phoneus* is now considered to be *Gonyaulax phoneus*.

Several different types of dinoflagellate poisons are now known to exist. They may vary in quantity and in type within a single species of dinoflagellate and the molluscs ingesting them.

The number of toxic dinoflagellates in the water varies according to the season of the year and the abundance of nutrient chemical substances that are available in the water. Over a period of time, it has been observed that the dangerous part of the year when dinoflagellates multiply is the warm season, which varies somewhat, but in north temperate regions is from March to November. An exception to this rule is the Alaskan butter clam, *Saxidomus giganteus* (Deshayes), which may have dangerous toxicity levels in certain coastal areas of Alaska at almost any time of the year.

SHELLFISH MOST COMMONLY INVOLVED IN HUMAN POISONING

Common Cockle, *Cardium edule* Linnaeus (Figure 88). Inhabits European Seas.

White Mussel, *Donax serra* (Chemnitz) (Figure 88). Inhabits South Africa.

Solid Surf Clam, *Spisula solidissima* (Dillwyn) (Figure 88). Occurs from Labrador to North Carolina.

Gaper or **Summer Clam,** *Schizothaerus nuttalli* (Conrad) (Figure 88). Found from Prince William Sound, Alaska, south to Scammons Lagoon, Baja California, and northern Japan.

Soft-Shelled Clam, *Mya arenaria* Linnaeus (Figure 89). Inhabits Britain, Scandinavia, Greenland, Atlantic coast of North America, south to Carolina; Alaska, south to Japan, and Vancouver, British Columbia; California and Oregon coasts.

Common Mussel, *Mytilus californianus* Conrad (Figure 89). Inhabits Unalaska, Aleutian Islands, eastward and southward to Socorro Island.

Bay Mussel, *Mytilus edulis* Linnaeus (Figure 89). Ranges from the Arctic Ocean to South Carolina, Alaska to Cape San Lucas, Baja California; practically worldwide in temperate waters.

Northern Horse Mussel, *Volsella modiolus* (Linnaeus) (Figure 89). Found along the Pacific Coast of America from the Arctic Ocean to San Ignacio Lagoon, Baja California; circumboreal.

Atlantic Jackknife or **Razor Clam,** *Ensis directus* Conrad (Figure 90). Ranges from the Gulf of St. Lawrence River to Florida.

Figure 88. A. **Common Cockle**, *Cardium edule* Linnaeus. B. **White mussel**, *Donax serra* (Chemnitz). C. **Solid Surf Clam,**

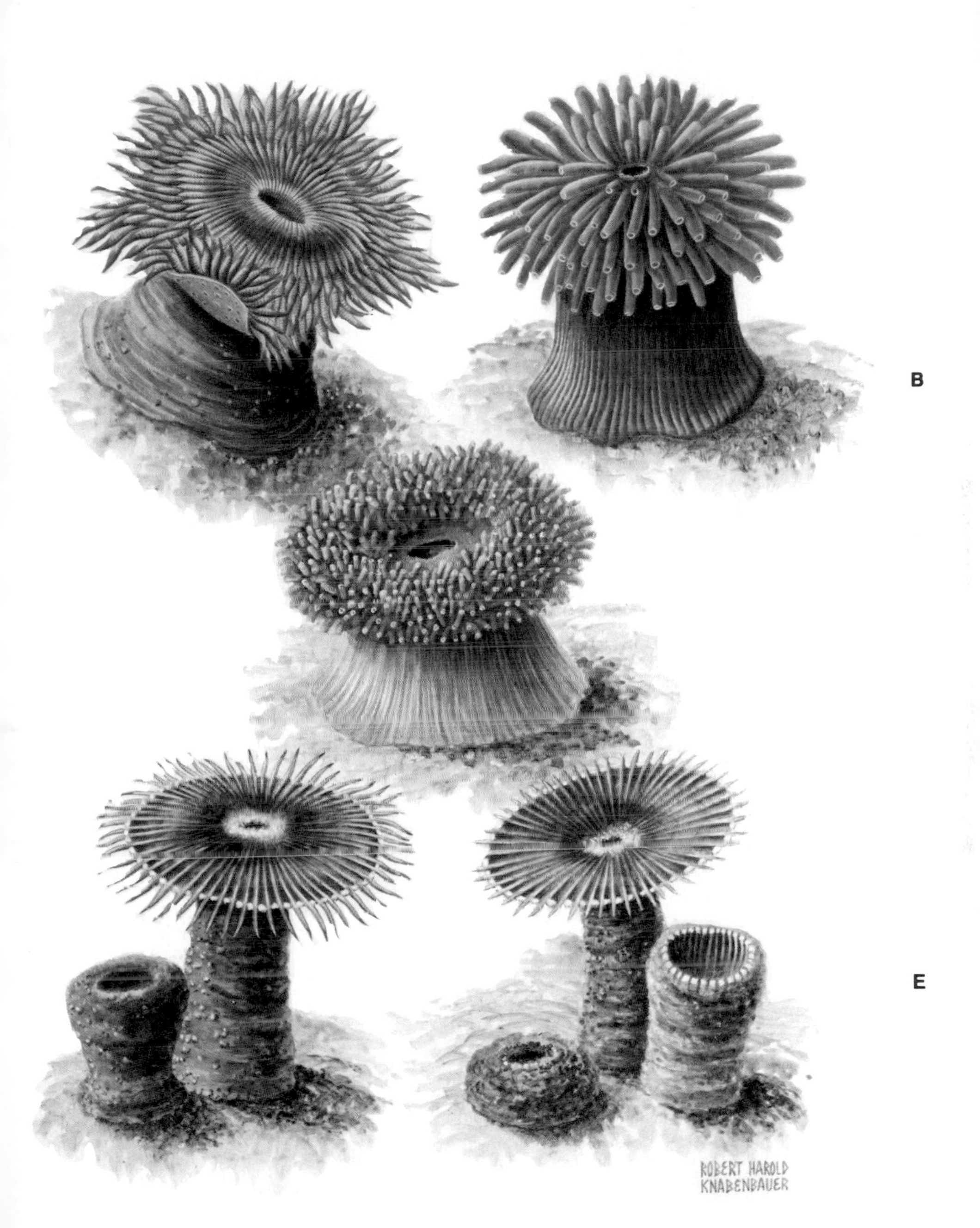

Plate 9. **Poisonous Sea Anemones:** A. *Radianthus paumotensis* (Dana). Diameter up to 3 inches (8 centimeters). B. *Physobrachia douglasi* Kent. Diameter up to 2 inches (5 centimeters). C. *Rhodactis howesi* Kent. Diameter about 2.5 inches (7 centimeters). D. *Palythoa toxica* Walsh and Bowers. Diameter of disk 3 inches (9 millimeters). E. *Palythoa tuberculosa* Esper. Diameter of disk 3 inches (9 millimeters).

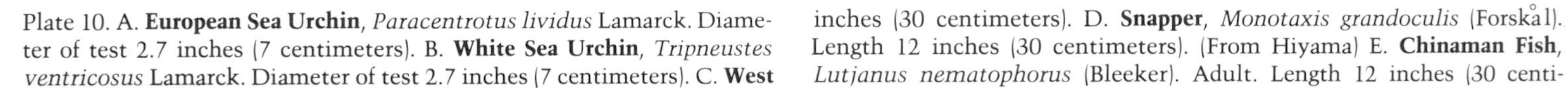

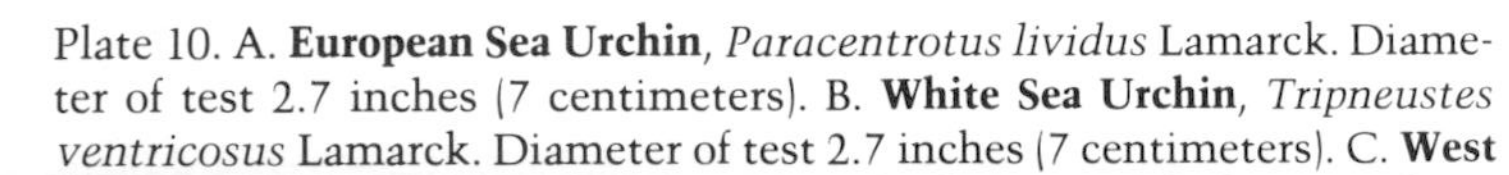

Plate 10. A. **European Sea Urchin**, *Paracentrotus lividus* Lamarck. Diameter of test 2.7 inches (7 centimeters). B. **White Sea Urchin**, *Tripneustes ventricosus* Lamarck. Diameter of test 2.7 inches (7 centimeters). C. **West** [illegible] inches (30 centimeters). D. **Snapper**, *Monotaxis grandoculis* (Forskål). Length 12 inches (30 centimeters). (From Hiyama) E. **Chinaman Fish**, *Lutjanus nematophorus* (Bleeker). Adult. Length 12 inches (30 centimeters).

A

B

C

D

Plate 11. A. **Blue Parrotfish**, *Scarus caeruleus* (Bloch). Length 35 inches (90 centimeters). B. **Parrotfish**, *Scarus microrhinos* Bleeker. Length 12 inches (30 centimeters). (From Hiyama) C. **Seabass** or **Grouper**, *Cephalopholis argus* Bloch and Schneider. Length 20 inches (50 centimeters). (From Hiyama) D. **Seabass**, *Plectropomus oligacanthus* Bleeker. Length 22 inches (56 centimeters). (From Hiyama)

A

B

C

D

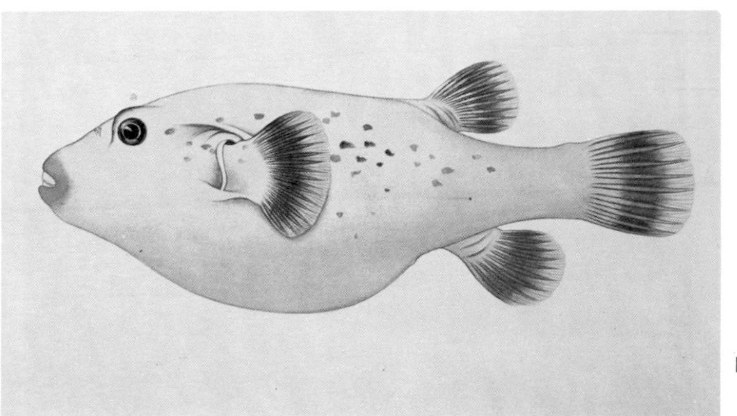

Plate 12. A. **Seabass**, *Plectropomus truncatus* (Fowler). Length 20 inches (50 centimeters). (From Hiyama) B. **White-spotted Puffer**, *Arothron meleagris* (Lacépède). Length 13 inches (33 centimeters). C. **Maki-maki**, or **Deadly Death Puffer**, *Arothron hispidus* (Linnaeus). Length 20 inches (50 centimeters). D. **Black-spotted Puffer**, *Arothron nigropunctatus* (Bloch and Schneider). Length 10 inches (25 centimeters).

A

B

C

D

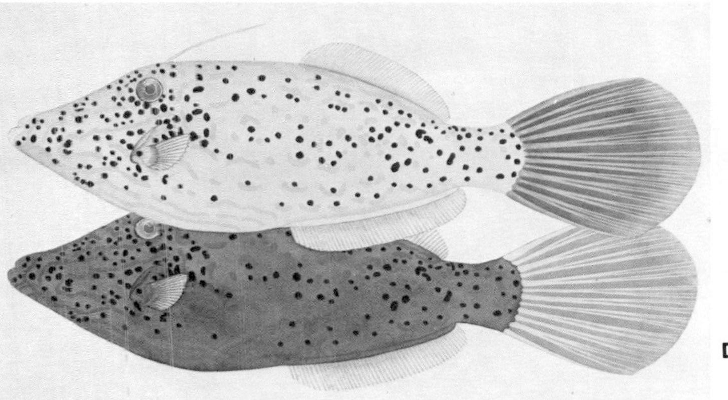

Plate 13. A. **Gulf Puffer**, *Sphaeroides annulatus* (Jenyns). Length 10 inches (25 centimeters). B. **Porcupine fish**, *Diodon hystrix* Linnaeus. Length 20 inches (50 centimeters). C. **Surgeonfish**, *Acanthurus triostegus* (Linnaeus). Length 8 inches (20 centimeters). (From Hiyama) D. **Filefish**, *Alutera scripta* (Osbeck). Length 20 inches (50 centimeters). (From Hiyama)

A

B

C

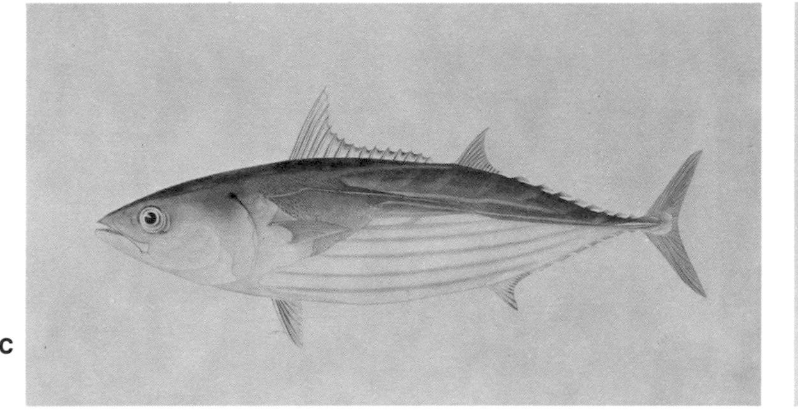

D

Plate 14. A. **Triggerfish**, *Balistoides conspicillum* Bloch and Schneider. Length 12 inches (30 centimeters). (From Hiyama) B. **Squirrelfish**, *Myripristis murdjan* (Forskål). Length 12 inches (30 centimeters). C. **Oceanic Bonito**, *Euthynnus pelamis* (Linnaeus). Length 20 inches (50 centimeters). D. **Wrasse**, *Epibulus insidiator* (Pallas). Length 12 inches (30 centimeters). (From Hiyama)

A

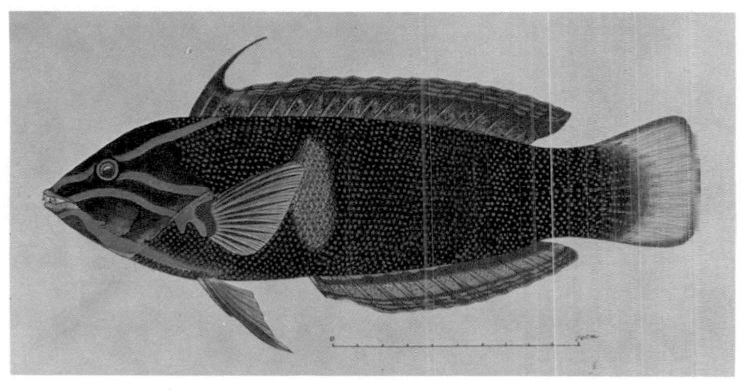

B

C

D

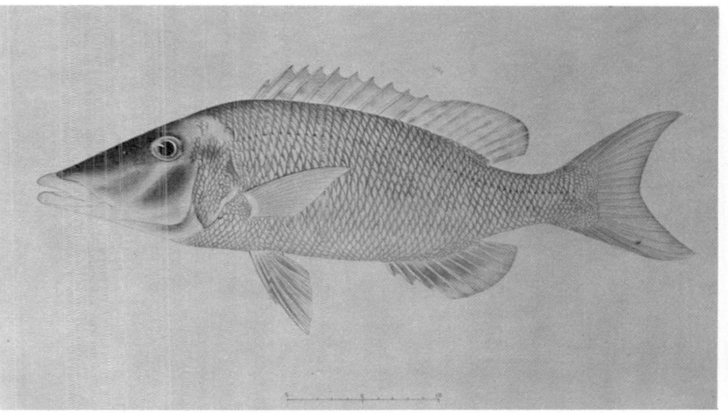

Plate 15. A. **Wrasse**, *Coris gaimardi* (Quoy and Gaimard). Length 12 inches (30 centimeters). B. **Snapper**, *Aprion virescens* Valenciennes. Length 27 inches (70 centimeters). (From Hiyama) C. **Snapper**, *Gnathodentex aureolineatus* (Lacepède). Length 10 inches (25 centimeters). (From Hiyama) D. **Snapper**, *Lethrinus miniatus* (Forster). Length 18 inches (45 centimeters). (From Hiyama)

A

B

C

D

Plate 16. A. **Snapper**, *Lutjanus bohar* (Forskål). Length 35 inches (90 centimeters). (From Hiyama) B. **Red Snapper**, *Lutjanus vaigiensis* (Quoy and Gaimard). Length 20 inches (50 centimeters). C. **Snapper**, *Lutjanus monostigma* (Cuvier). Length 12 inches (30 centimeters). (From Hiyama) D. **Snapper**, *Lutjanus gibbus* (Forskål). Length 15 inches (40 centimeters). (From Hiyama)

Figure 89. A. **Soft-shelled Clam,** *Mya arenaria* Linnaeus. B. **Common Mussel,** *Mytilus californianus* Conrad. C. **Bay Mussel,** *Mytilus edulis* Linnaeus. D. **Northern Horse Mussel,** *Volsella modiolus* (Linnaeus).

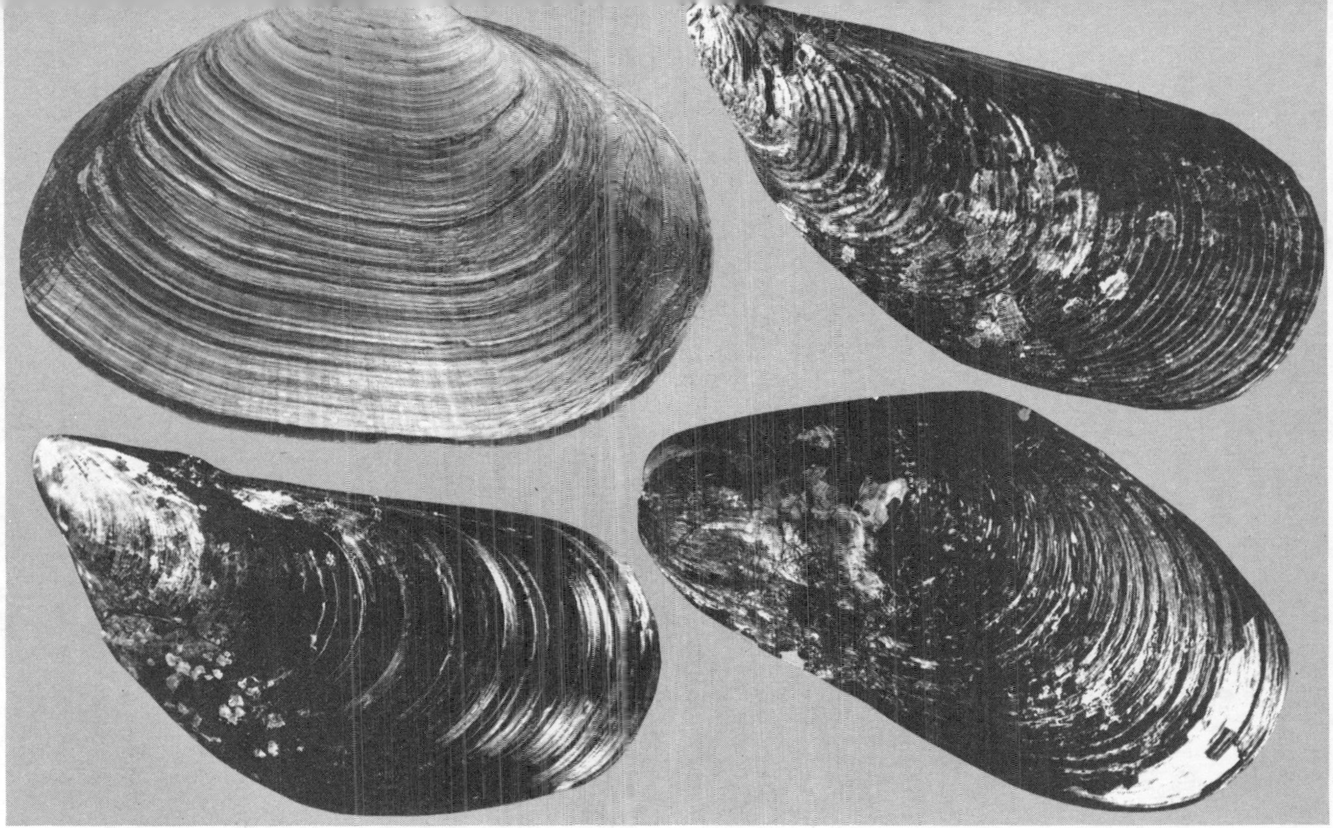

Alaskan Butter Clam, Smooth Washington or **Butter Clam,** *Saxidomus giganteus* (Deshayes) (Figure 90). Ranges from Sitka, Alaska, to San Francisco Bay, California.

Common Washington or **Butter Clam,** *Saxidomus nuttalli* Conrad (Figure 90). Ranges from Humbolt Bay, California to San Quentin Bay, Baja California.

Medical Aspects: Three types of shellfish poisoning are recognized by physicians:

1. *Gastrointestinal type*—characterized by such symptoms as nausea, vomiting, diarrhea, and abdominal pain. This type usually develops about 10-12 hours after eating the shellfish, and is believed to be caused by bacterial contamination;

2. *Allergic type*—characterized by redness of the skin, swelling, development of a hivelike rash, itching, headache, nasal congestion, abdominal pain, dryness of the throat, swelling of the tongue, palpitation of the heart, and difficulty in breathing. This type probably results from a sensitivity to shellfish on the part of the individual;

3. *Paralytic type*—this last type is caused specifically by the dinoflagellate poison present in shellfish. The disease has also been termed paralytic shellfish (PSP), clam, mussel, or gonyaulax poisoning. The early symptoms are a tingling or burning sensation of the lips, gums, tongue, and face, which gradually spreads elsewhere to the body. The tingling areas later become numb, and movements of the muscles of the body may become very difficult. Other symptoms frequently present are weakness, dizziness, joint aches, increased salivation, intense thirst, difficulty in swallowing. Nausea, vomiting, diarrhea, and abdominal pain are relatively rare. The muscular paralysis may become increasingly severe until death ensues.

Inhalation of toxic products contained in windblown spray from red tide areas of *Gymnodinium breve* may cause irritation of the mucous membranes of the nose and throat, resulting in coughing, sneezing, and respiratory distress.

Treatment: There is no specific treatment available for paralytic shellfish poisoning and no known antidotes. Evacuation of the gastrointestinal tract should be instituted as soon as possible if shellfish have

Figure 90. A. **Atlantic Jacknife** or **Razor Clam,** *Ensis directus* Conrad. B. **Smooth Washington** or **Butter Clam,** *Saxidomus giganteus* (Deshayes). C. **Common Washington** or **Butter Clam,** *Saxidomus nuttalli* Conrad. ▸

A
B
C

A

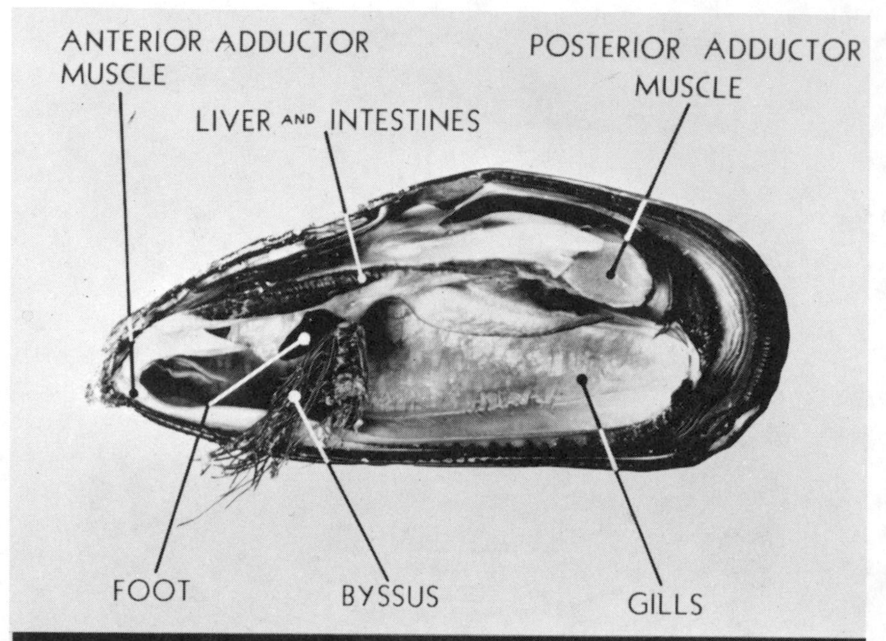
ANTERIOR ADDUCTOR
MUSCLE
LIVER AND INTESTINES
POSTERIOR ADDUCTOR
MUSCLE
FOOT
BYSSUS
GILLS

B

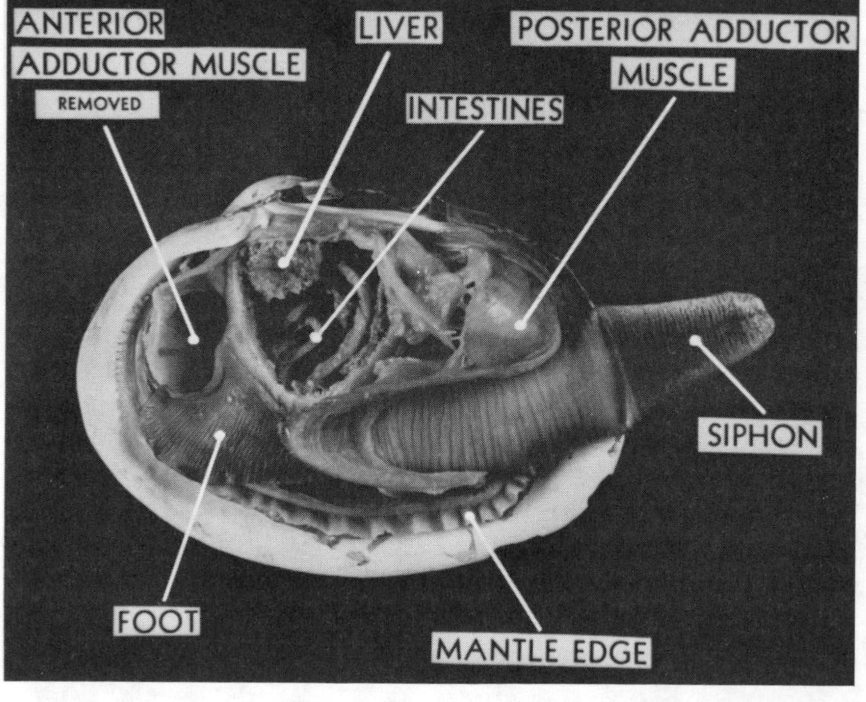
ANTERIOR
ADDUCTOR MUSCLE
REMOVED
LIVER
POSTERIOR ADDUCTOR
MUSCLE
INTESTINES
SIPHON
FOOT
MANTLE EDGE

been ingested. Vomiting can be stimulated by swallowing large quantities of salt water, egg white, or by merely placing one's finger down the throat. Alkaline fluids as a solution of ordinary baking soda are said to be of value since the poison is rapidly destroyed by these fluids. Artificial respiration may be required. See a physician at once if you are fortunate enough to be near one.

Prevention: The extremely toxic nature of this poison cannot be over-emphasized. Most areas where paralytic shellfish poisoning is likely to occur are examined by local public health authorities. When toxic shellfish are discovered, the area is placed under quarantine. One should adhere strictly to local quarantine regulations. Since poisonous shellfish cannot be detected by their appearance, smell, or by discoloration of a silver object or garlic placed in cooking water, etc., it is only by careful scientific laboratory procedures that paralytic shellfish poison can be determined with any degree of certainty.

The digestive organs, or dark meat, gills, and in some shellfish species the siphon, contain the greatest concentration of the poison (Figure 91). The musculature or white meat is generally harmless; however, it should be thoroughly washed before cooking. The broth, or bouillon, in which the shellfish is boiled is especially dangerous since the poison is water soluble. The broth should be discarded if there is the slightest doubt. The tidal location from which the shellfish were gathered cannot be used as a criterion for whether the shellfish are safe to eat. Poisonous shellfish may be found in either low or high tidal zones. If in doubt—throw them out!

POISONOUS COELENTERATES

Poisonous Sea Anemones. Intoxications resulting from the ingestion of poisonous sea anemones are extremely rare. However, poisonings have occurred in Samoa and other parts of the tropical Indo-Pacific region. Sea anemones are commonly eaten by the natives of Samoa, but only certain species, and, only after the anemones are cooked. *Rhodactis howesi* Kent (Plate 9), known locally as *matalelei*, and *Physobrachia*

◂ Figure 91. Internal anatomy of shellfish, showing parts of the body likely to contain paralytic shellfish poison. A. **Mussel,** *Mytilus.* Liver and intestines (dark meat) and gills are the most dangerous parts to eat. B. **Butter Clam,** *Saxidomus.* The poison is concentrated primarily in the siphon.

douglasi Kent (Plate 9), known locally as *lumane*, are generally considered to be poisonous when raw, but safe to eat when cooked. *Radianthus paumotensis* (Dana) (Plate 9), known locally as *matamala samasama*, is considered to be poisonous either raw or cooked.

Medical Aspects: Initial symptoms are those of acute gastritis including nausea, vomiting, abdominal pain, cyanosis, and prostration. Shortly after ingestion the victim may become stuporous with an absence of superficial reflexes, but with a normal pulse rate and blood pressure. Eventually the patient may go into prolonged shock and die with a pulmonary edema.

Treatment: Symptomatic, no known specific antidote.

Poisonous Zoanthids. Some of the zoanthids, or soft-corals, of the genus *Palythoa* have been found to contain one of the most deadly poisons known to marine biotoxicologists. This poison is produced within the body of the organism rather than in the stinging nematocysts of the tentacles. The poison, designated "palytoxin," has been demonstrated to have a potent action on the nervous and muscular systems of experimental animals (crabs and mice), but little is known concerning the effects of palytoxin in humans. The poison can be inactivated with heat and acids, and is probably destroyed by gastric juices. When swimming in closed tidal pools containing large numbers of *Palythoa* a person may develop sensations of numbness and tingling of the lips and mouth.

Species of the soft-coral *Palythoa* are found both in the Caribbean and the tropical Pacific. In the Hawaiian Islands *Palythoa* is referred to as the "deadly seaweed of Hana (Limu-Make O Hana)." *Palythoa* (Plate 9) are generally found growing in tidal pools or other shallow water protected area, and may be encrusted on shells, seaweed, corals, or rocks.

Persons working for prolonged periods in confined tidal pools inhabited by this coelenterate should do so with caution. One should keep in mind that palytoxin is extremely poisonous and may cause serious skin reactions to sensitive persons. Contact with the slime of *Palythoa* in an open wound may cause malaise, muscle cramps, abdominal pain, and could be fatal. Handle this sea anemone with extreme care.

POISONOUS ECHINODERMS

There are relatively few echinoderms (starfishes, sea urchins, sea cucumbers) known to produce human oral intoxications. Only two groups have been incriminated to any significant extent, namely:

Sea Urchin Eggs. Reports have appeared from time to time of poisonings due to the ingestion of sea urchin eggs. The **European Sea Urchin,** *Paracentrotus lividus* Lamarck (Plate 10), inhabiting the Atlantic coasts of Europe and the Mediterranean, and the **White Sea Urchin,** *Tripneustes ventricosus* (Lamarck) (Plate 10), inhabiting the West Indies, have been incriminated. It is believed that the poison is contracted from algae upon which the sea urchins have been feeding.

Figure 92. **Asiatic Horseshoe Crab,** *Carcinoscorpius rotundicauda* (Latreille). Length 13 inches (33 centimeters).

Poisonings usually occur during the reproductive season of the year. Symptoms consist of an acute gastritis, nausea, vomiting, diarrhea, abdominal pain, and severe migrainelike headaches. Little else is known about sea urchin poisoning in humans, or the nature of the poison.

Sea Cucumbers. These are a group of sluggish creatures which move over the bottom of the sea by means of rhythmic contractions of a sausagelike body. They have a series of tentacles circling the mouth at the anterior end of the body. Sea cucumbers, members of the class *Holothuroidea,* serve as important items of food in some parts of the world where they are sold under the name of "Trepang" or "beche-de-mer." They are boiled and then dried in the sun or smoked. "Trepang" is used to flavor soups and stews.

Contact with liquid ejected from the visceral cavity of some sea cucumber species may result in dermatitis or blindness. The poison of sea cucumbers is termed "holothurin" and is generally concentrated in the organs of Cuvier. Intoxications may also occur as a result of ingestion of sea cucumbers, but poisonings of this type are rare. Nothing is known concerning the symptoms produced, but fatalities have been reported.

Treatment of sea cucumber poisoning is symptomatic. There are no known antidotes.

POISONOUS ARTHROPODS

The phylum *Arthropoda* is the largest single group of the animal kingdom, having more than 800,000 species. Unfortunately, very little is known concerning the toxicity of most marine arthropods, the joint-legged animals which includes the horseshoe crabs, lobsters, and crabs. Some of the Asiatic horseshoe crabs, tropical lobsters, and crabs are occasionally quite poisonous to eat.

ASIATIC HORSESHOE CRABS

Poisoning from ingestion of the **Asiatic Horseshoe Crab,** *Carcinoscorpius rotundicauda* (Latreille) (Figure 92), is due to the eating of the unlaid green eggs or viscera during the reproductive season of the year. Most of the intoxications have occurred in southeast Asia. Despite their periodic toxicity, the large masses of green unlaid eggs are highly esteemed by Asiatic peoples.

Medical Aspects: Horseshoe crab poisoning, or mimi poisoning, usually occurs within 30 minutes after ingestion of the toxic material. Symptoms consist of dizziness, headaches, nausea, vomiting, abdominal cramps, diarrhea, cardiac palpitation, numbness of the lips, tingling of the lower extremities, weakness, loss of speech, sensation of heat in the mouth, throat and stomach, muscular paralysis, hypersalivation, drowsiness, and loss of consciousness. The mortality rate is very high. Death may occur within 16 hours or less.

Treatment: The treatment is symptomatic. There is no known antidote. Use the same basic procedure as in ciguatera or puffer fish poisoning.

Prevention: Although Asiatic horseshoe crabs are eaten in many parts of southeast Asia, they should be avoided during the reproductive season.

TROPICAL REEF CRABS

Although violent intoxications have been periodically reported from eating tropical reef crabs, it is only within the last few years that serious scientific effort has been made to study the occurrence and nature of crab poisons.

There are numerous crab species which occur in tropical waters, but only a few species have been found to be toxic. One of the toxic species is the **Coconut Crab,** *Birgus latro* (Linnaeus) (Figure 93), which is terrestrial and found in damp jungle areas on tropical islands of the Indo-Pacific. Intoxications from the coconut crab have been reported in the Ryukyu Islands. It is believed this crab becomes toxic as a result of feeding on poisonous plants. The symptoms of coconut crab poisoning include nausea, vomiting, headache, chills, fever, joint aches, exhaustion, and muscular weakness. Deaths have been reported.

Several species of tropical reef crabs have produced violent intoxications. They include the following:

Reef Crab, *Atergatus floridus* (Linnaeus) (Figure 93). Indo-Pacific.

Red Spotted Crab, *Carpilius maculatus* Linnaeus (Figure 93). Indo-Pacific.

Reef Crab, *Demania toxica* Garth (Figure 93). Indo-Pacific.

Medical Aspects: The symptoms produced by tropical reef crabs differ from those of the coconut crab. The symptoms include nausea, vomiting, collapse, numbness, tingling, and muscular paralysis. Death may occur in severe intoxications.

The poison is believed to be chemically identical to tetrodotoxin, the causative agent of puffer poison.

Treatment: Symptomatic, no known antidote.

Prevention: Avoid eating tropical reef crabs unless you are certain they are safe to eat. If in doubt, check with the local natives or public health authorities.

B

Figure 93. A. **Reef Crab,** *Atergatus floridus* (Linnaeus). Width of carapace 2 inches (5 centimeters). B. **Red Spotted Crab,** *Carpilius maculatus* Linnaeus. Width of carapace 5 inches (13 centimeters). C. **Reef Crab,** *Demania toxica* Garth. Width of carapace 2 inches (5 centimeters). D. **Coconut Crab,** *Birgus latro* (Linnaeus).

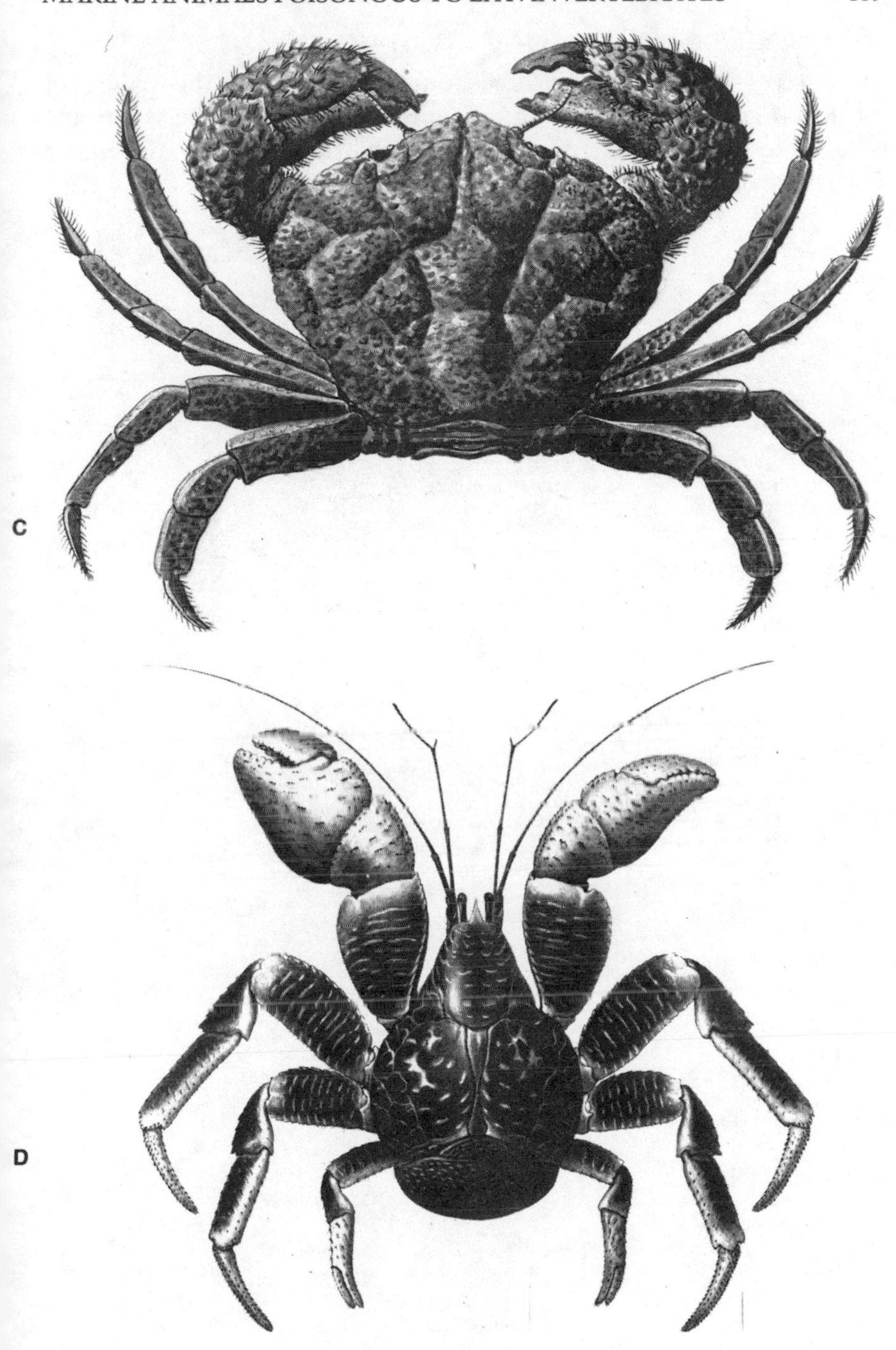
C
D

LOBSTERS

Several reports have appeared from time to time on the toxicity of lobsters (*Panulirus spider*) found in southern Polynesia. The symptoms seem to resemble ciguatera fish poison but little is known concerning the nature of the poison.

VI

MARINE ANIMALS POISONOUS TO EAT: VERTEBRATES

POISONOUS FISHES

Tropical reef fishes may contain some of the most deadly poisons known to marine biotoxicologists. As previously indicated there are a variety of forms of ichthyosarcotoxism, or fish-flesh-poisoning. The following is a brief resume of some of the more common types.

ELASMOBRANCH POISONING

This form of ichthyosarcotoxism is most commonly caused by eating the liver of sharks. However, the flesh of some of the larger tropical sharks and the Greenland shark, *Somniosus microcephalus,* which inhabits Arctic waters, has caused intoxications in humans and sled dogs. The chemical nature of these shark poisons is unknown.

SPECIES REPORTED POISONOUS

Black-Tipped Reef Shark, *Carcharhinus melanopterus* (Quoy and Gaimard) (Figure 94). Indo-Pacific, South Africa to the East Indies, Hawaiian, Tuamotu, and Marianas Islands.

Greenland Shark, *Somniosus microcephalus* (Bloch and Schneider) (Figure 94). Arctic Atlantic, North Sea east to the White Sea and west to the Gulf of St. Lawrence, Greenland.

Hammerhead Shark, *Sphyrna zygaena* (Linnaeus) (Figure 9). Tropical to warm-temperate belt of the Atlantic and Pacific Oceans.

Seven-Gilled Shark, *Heptranchias perlo* (Bonnaterre) (Figure 94). East and west Atlantic, Mediterranean, Cape of Good Hope, Japan.

Six-Gilled Shark, *Hexanchus grisseus* (Bonnaterre) (Figure 94). Atlantic, Pacific coast of North America, Chile, Japan, Australia, Southern Indian Ocean and South Africa.

Great White Shark, *Carcharodon carcharias* (Linnaeus) (Plate 2). Cosmopolitan, in tropical, subtropical and warm-temperate belts of all oceans.

A
B
C
D

Medical Aspects: The most severe forms of poisoning usually result from the eating of the liver. The musculature in most instances is only mildly toxic with the symptoms seldom more than that of a mild gastrointestinal upset and a predominating diarrhea. Symptoms from liver poisoning usually develop within 30 minutes, and consist of nausea, vomiting, diarrhea, abdominal pain, headache, joint aches, tingling about the mouth, and a burning sensation of the tongue, throat, and esophagus. As time goes on, the nervous symptoms may become progressively severe, resulting in muscular incoordination and difficulty in breathing due to muscular paralysis, coma, and finally death.

Treatment: See treatment of fish poisoning.

Prevention: Avoid eating the liver of any shark unless it is known with certainty to be edible. The livers of large tropical sharks are said to be especially dangerous. The flesh of tropical and arctic sharks should be indulged in only with caution.

CIGUATERA

This is one of the serious and widespread forms of ichthyosarcotoxism. Biotoxication is caused largely by tropical shorefishes, with more than 400 species incriminated. Because many of the toxic species are generally regarded as valuable food, ciguatoxic fish constitute a serious threat to the development of tropical shore-fisheries. In many instances, useful food fishes suddenly, and without warning, become poisonous within a matter of hours and may remain toxic for a period of years, and it is believed that they do so because they have been feeding upon some noxious material such as toxic algae, invertebrates, fishes, or possibly dinoflagellates. Figure 95 is a diagram depicting the biogenesis or transvectoring mechanism by which ciguatoxin is obtained through the food chain. There is now evidence that a dinoflagellate (*Gambierdiscus toxicus* Adachi and Fukuyo) which is frequently found on the surface of a benthic brown seaweed (*Turbinaria ornata* J. Agardh) and other algae species may produce ciguatoxin. This dinoflagellate, and possibly other

◂ Figure 94 A. **Black-Tipped Reef Shark,** *Carcharhinus melanopterus* (Quoy and Gaimard). Length 6.3 feet (2 meters). B. **Greenland Shark,** *Somniosus microcephalus* (Bloch and Schneider). Length 6.3 feet (2 meters). C. **Seven-Gilled Shark,** *Heptranchias perlo* (Bonnaterre). Length 6.3 feet (2 meters). D. **Six-Gilled Shark,** *Hexanchus grisseus* (Bonnaterre). Length 15.5 feet (5 meters). The liver or flesh of these sharks may be poisonous to eat.

species as well, may serve as the primary source of ciguatoxin. These dinoflagellates may be ingested by filter feeding invertebrates, plankton feeding fishes, herbivorous fishes feeding on marine plants, and indirectly by carnivorous fishes feeding on herbivorous fishes. It is evident that the source of ciguatoxin is due to a complex food web involvement in which a variety of environmental factors are playing a role. The food web interactions account for the widespread phylogenetic distribution of ciguatoxin among marine animals.Ciguatera is most prevalent in subtropical and tropical latitudes, but the greatest concentration of ciguatoxic fishes seems to be around the islands of the tropical Pacific, Caribbean, and Indian Oceans.

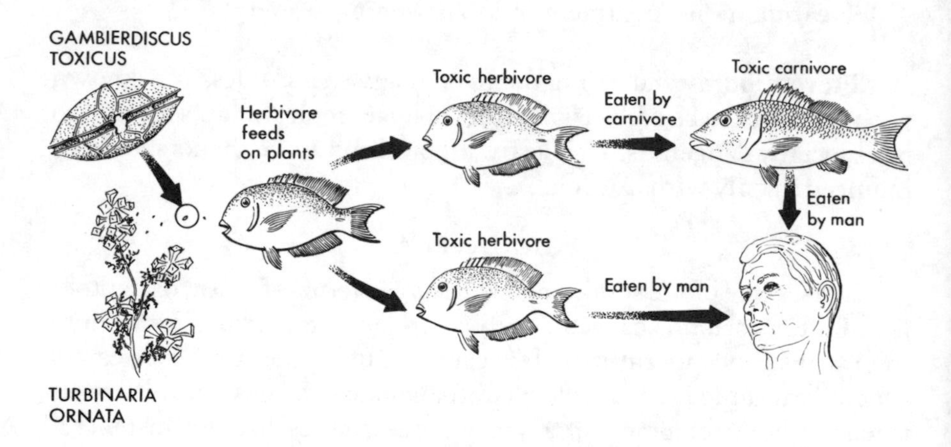

Figure 95. Diagram depicting the biogenesis or transvectoring mechanism by which ciguatoxin is obtained through the food chain.

REPRESENTATIVE SPECIES OF CIGUATERA-PRODUCING FISHES

Anchovy, *Engraulis japonicus* (Schlegel) (Figure 96). China, Japan, Korea, Formosa.

Barracuda, *Sphyraena barracuda* (Walbaum) (Figure 96). Indo-Pacific; from Hawaii to the Red Sea; west Atlantic from Brazil to the West Indies, Florida and Bermuda.

Chinaman Fish, *Lutjanus nematophorus* (Bleeker) (Plate 10). Australia.

Filefish, *Alutera scripta* (Osbeck) (Plate 13). All warm seas.

Figure 96. A. **Anchovy,** *Engraulis japonicus* (Schlegel). Length 5 inches (13 centimeters). B. **Barracuda,** *Sphyraena barracuda* (Walbum). Length 5 feet (1.5 meters). C. **Herring,** *Clupanodon thrissa* (Linnaeus). Length 10 inches (25 centimeters). D. **Jack,** *Caranx hippos* (Linnaeus). Length 25 inches (65 centimeters). E. **Jack,** *Caranx melampygus* (Cuvier). ▸

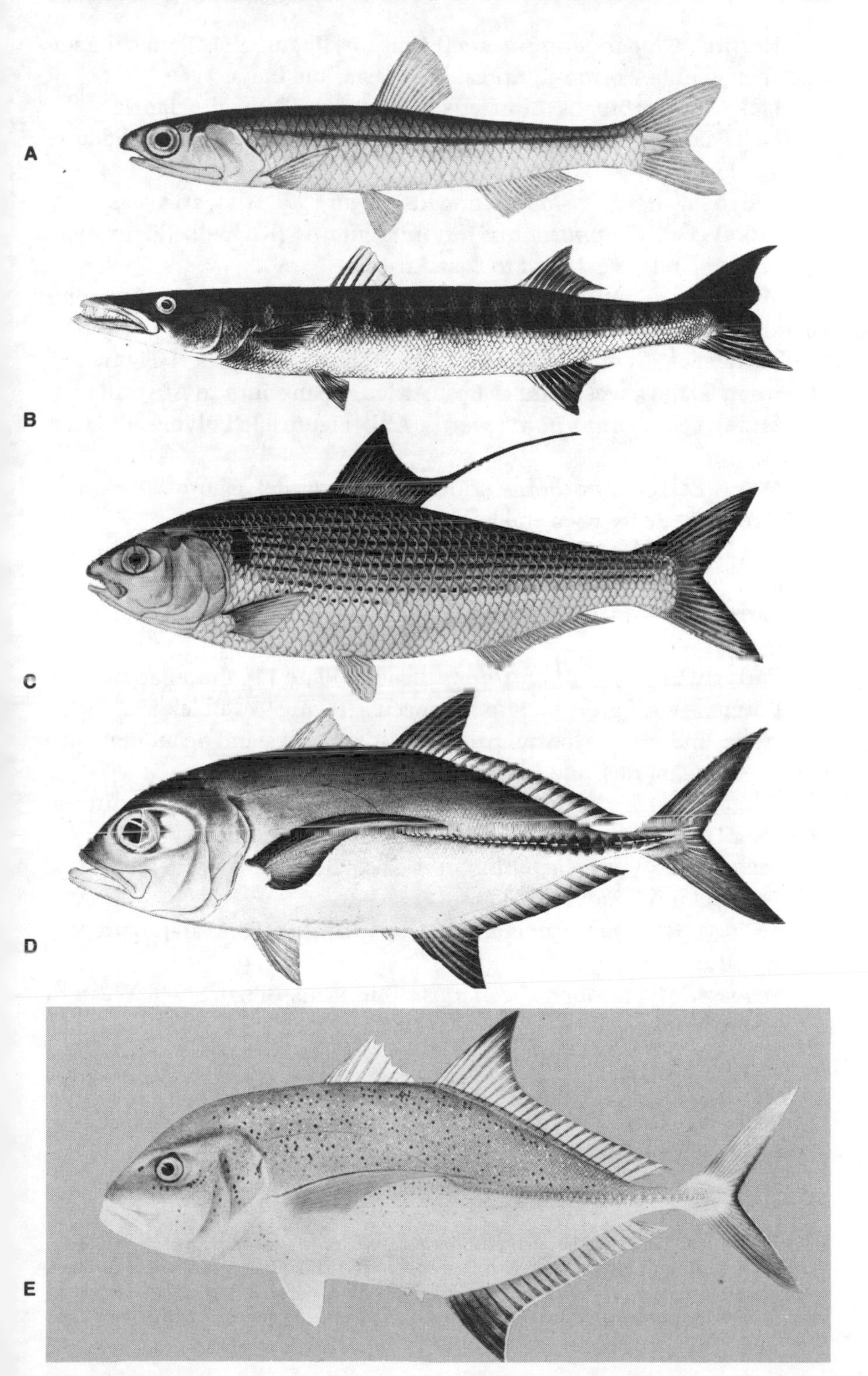
A
B
C
D
E

Herring, *Clupanodon thrissa* (Linnaeus) (Figure 96). Tropical Pacific, Japan, China, Formosa, Korea, Indonesia, and India.

Jack, *Caranx hippos* (Linnaeus) (Figure 96). Tropical Atlantic.

Jack, *Caranx melampygus* (Cuvier) (Figure 96). Tropical Pacific, north to Japan.

Ladyfish, *Albula vulpes* (Linnaeus) (Figure 97). All warm seas.

Moray Eel, *Gymnothorax flavimarginatus* (Rüppell) (Figure 98). Hawaiian Islands, westward to East Africa.

Moray Eel, *Gymnothorax javanicus* (Bleeker) (Figure 98). Hawaiian Islands, westward to East Africa.

Moray Eel, *Gymnothorax meleagris* (Shaw and Nodder) (Figure 98). Hawaiian Islands, westward to East Africa, Japan south to Australia.

Moray Eel, *Gymnothorax pictus* (Ahl) (Figure 98). Polynesia to East Africa.

Moray Eel, *Gymnothorax undulatus* (Lacepede) (Figure 98). Hawaiian Islands to the Red Sea and East Africa.

Oceanic Bonito, *Euthynnus pelamis* (Linnaeus) (Plate 14). Circumtropical.

Parrotfish, *Scarus caeruleus* (Bloch) (Plate 11). Florida and West Indies.

Parrotfish, *Scarus microrhinos* Bleeker (Plate 11). Indo-Pacific.

Porgie, *Pagellus erythrinus* (Linnaeus) (Figure 97). Black Sea, Mediterranean, and east Atlantic, from the British Isles and Scandinavia to the Azores, Canaries, and Fernando Po.

Porgie, *Pagrus pagrus* (Linnaeus). Eastern Atlantic and Mediterranean Sea.

Seabass, Grouper, *Cephalopholis argus* Bloch and Schneider (Plate 11). Tropical Indo-Pacific.

Seabass, Grouper, *Epinephelus fuscoguttatus* (Forskål) (Figure 97). Indo-Pacific.

Seabass, *Mycteroperca venenosa* (Linnaeus) (Figure 97). Western tropical Atlantic.

Seabass, *Plectropomus obligacanthus* Bleeker (Plate 11). Indonesia, Philippine, Caroline and Marshall Islands.

Seabass, *Plectropomus truncatus* (Fowler) (Plate 12). Micronesia, Indonesia, Philippines.

Seabass, *Variola louti* (Forskål) (Figure 97). Tropical Indo-Pacific.

Figure 97. A. **Ladyfish,** *Albula vulpes* (Linnaeus). Length 39.4 inches (1 meter). B. **Porgie,** *Pagellus erythrinus* (Linnaeus). Length 16 inches (40 centimeters). C. **Seabass** or **Grouper,** *Epinephelus fuscoguttatus* (Forskål). D. **Seabass,** *Mycteroperca venenosa* (Linnaeus). Length 35 inches (90 centimeters). E. **Seabass,** *Variola louti* (Forskål). Length 24 inches (60 centimeters). ▸

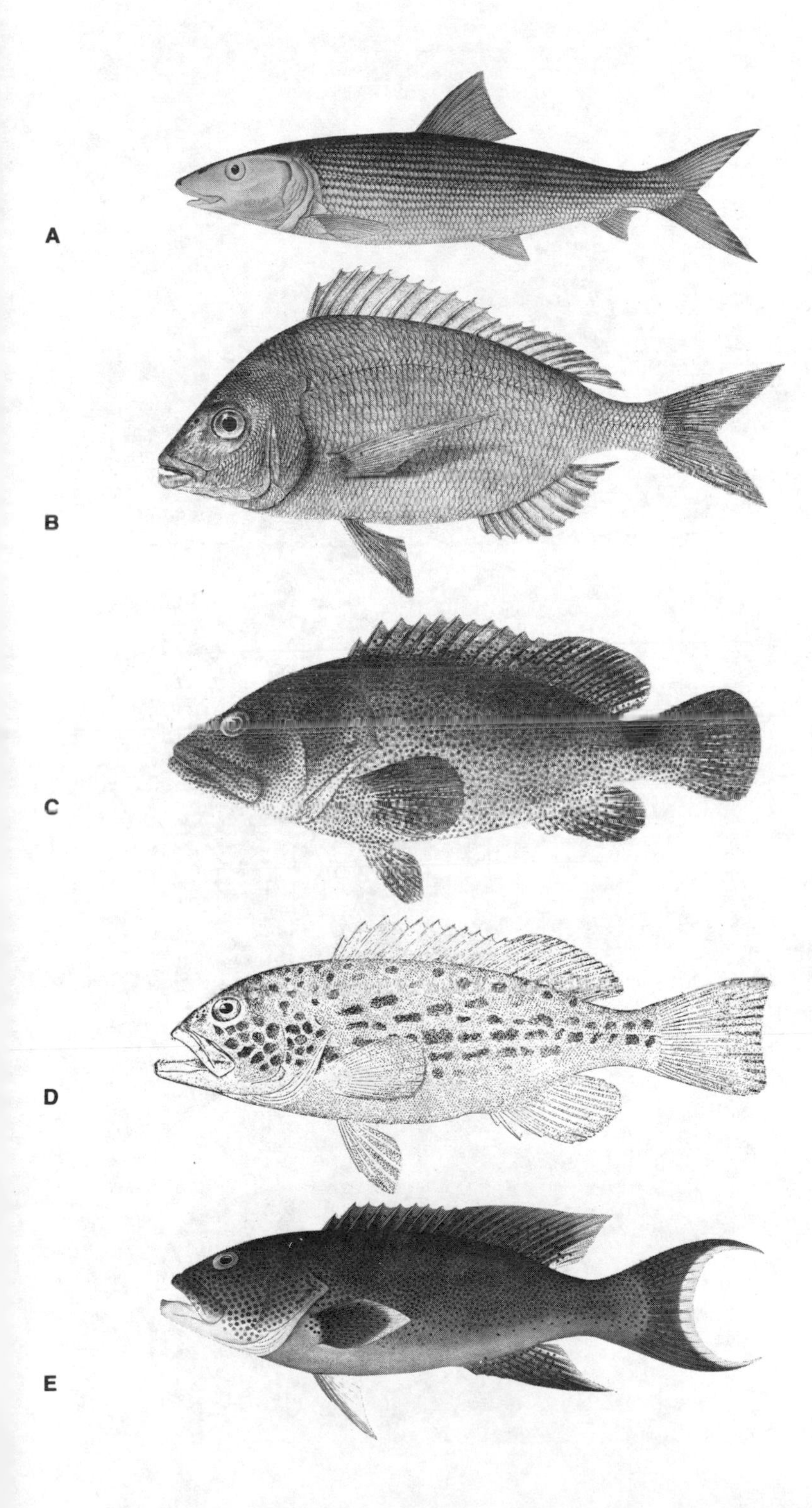
A
B
C
D
E

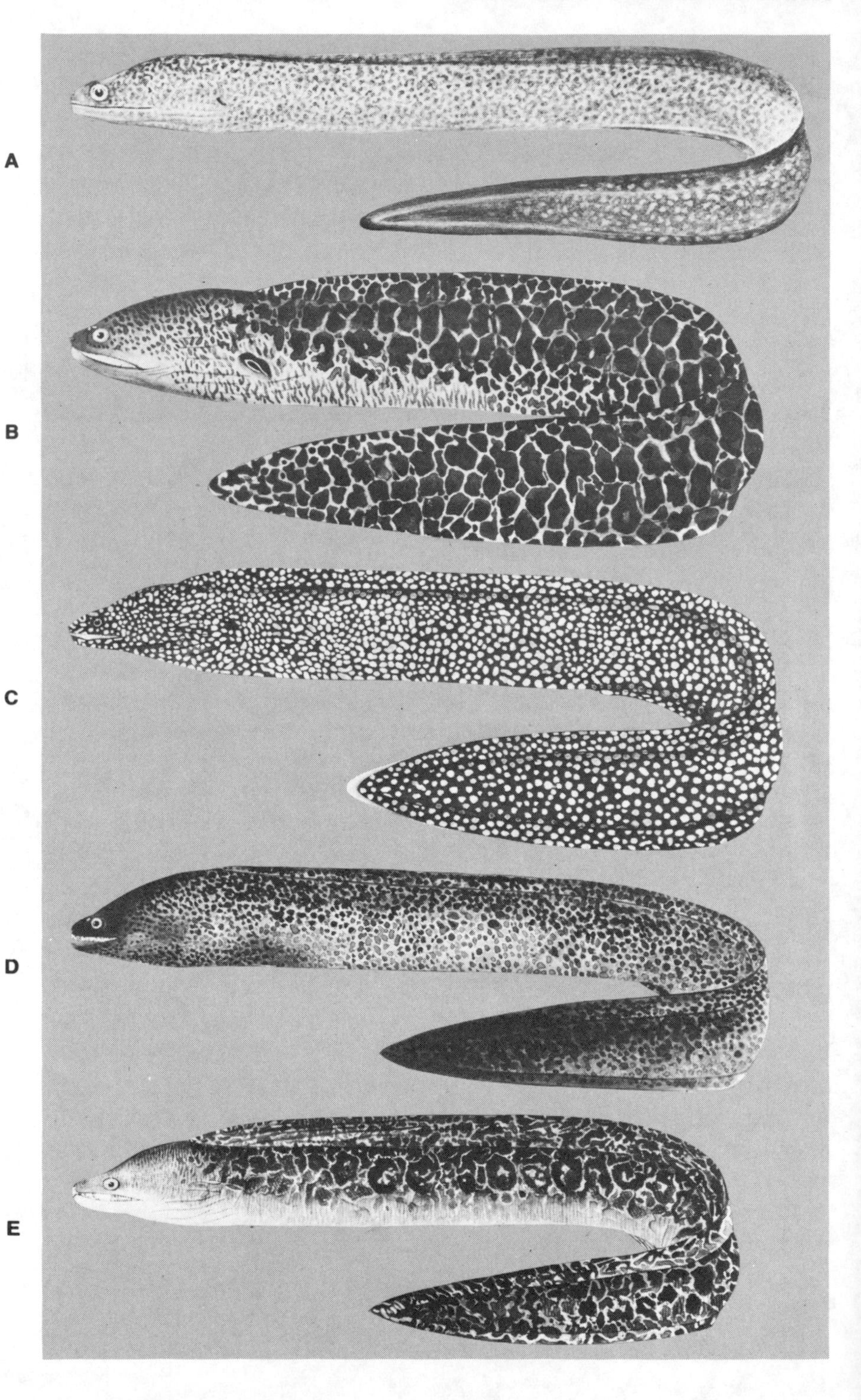
A
B
C
D
E

Snapper, Red, *Lutjanus bohar* (Forskål) (Plate 16). Tropical Pacific to east Africa and Red Sea.

Snapper, Red, *Lutjanus gibbus* (Forskål) (Plate 16). Tropical Indo-Pacific.

Snapper, Red, *Lutjanus vaigiensis* (Quoy and Gaimard) (Plate 16). Polynesia, westward to east Africa, Japan.

Snapper, *Aprion virescens* Valenciennes (Plate 15). Tropical Indo-Pacific.

Snapper, *Gnathodentex aureolineatus* (Lacépède) (Plate 15). Tuamotu Archipelago westward to East Africa.

Snapper, *Lethrinus miniatus* (Forster) (Plate 15). Polynesia, westward to East Africa.

Snapper, *Lutjanus monostigma* (Cuvier) (Plate 16). Polynesia, westward to the Red Sea, China.

Snapper, *Monotaxis grandoculis* (Forskål) (Plate 10). Polynesia, westward to east Africa.

Squaretail, *Tetragonurus cuvieri* (Risso) (Figure 99). Temperate regions of the world.

Squirrelfish, *Myripristis murdjan* (Forskål) (Plate 14). Indo-Pacific.

Surgeonfish, *Acanthurus glaucoparcius* (Cuvier) (Figure 99). Indonesia, Philippine Islands, and tropical Pacific.

Surgeonfish, *Acanthurus triostegus* (Linnaeus) (Plate 12). Hawaiian and Johnston Islands.

Surmullet, Goatfish, *Parupeneus chryserydros* (Lacépède) (Figure 99). Polynesia westward to east Africa.

Surmullet, Goatfish, *Upeneus arge* (Jordan and Evermann) (Figure 99). Polynesia and Micronesia.

Triggerfish, *Balistoides conspicillum* Bloch and Schneider (Plate 14). Tropical Pacific, from Polynesia to Madagascar, China, and Japan.

Trunkfish, *Lactophrys trigonus* (Linnaeus) (Figure 100). Atlantic coast of tropical America, northward to Cape Cod.

Trunkfish, *Lactoria cornutus* (Linnaeus) (Figure 100). Tropical Pacific.

Wrasse, *Epibulus insidiator* (Pallas) (Plate 14). Tropical Indo-Pacific.

Wrasse, *Coris gaimardi* (Quoy and Gaimard) (Plate 15). Tropical Indo-Pacific area.

◂ Figure 98. **Moray Eel,** *Gymnothorax flavimarginatus* (Rüppell). Length 60 inches (1.5 meters). **Moray Eel,** *Gymnothorax javanicus* (Bleeker). **Moray Eel,** *Gymnothorax meleagris* (Shaw and Nodder). Length 39.4 inches (1 meter). **Moray Eel,** *Gymnothorax pictus* (Ahl). Length 28 inches (75 centimeters). **Moray Eel,** *Gymnothorax undulatus* (Lacépède). Length 60 inches (1.5 centimeters).

Medical Aspects: Symptoms of ciguatera poisoning consist of paraesthesias of the lips, tongue and limbs, and gastrointestinal disturbances. Victims frequently complain of myalgia, joint aches and profound muscular weakness. The so-called paradoxical sensory disturbance in which the victim interprets cold as a "tingling, burning, dry-ice, or electric-shock sensation," or hot objects as cold, is characteristic of this form of poisoning. Severe neurological disturbances consisting of ataxia, generalized motor incoordination, diminished reflexes, muscular twitching, tremors, dysponia, dysphagia, clonic and tonic convulsions, coma and muscular paralysis may be present. The case fatality rate is said to be about 12 percent. Complete recovery may require many months, and even years in some cases.

Clupeoid fishes (anchovies, herrings, etc.) have been known to cause violent poisonings at sporadic intervals in the tropical Atlantic, Caribbean Sea, tropical Pacific, and Indian Ocean. Most poisonings have occurred in the vicinity of tropical islands. Tropical clupeiform fishes are most likely to be toxic during the warm summer months. The first indication of poisoning is a sharp metallic taste which may be present immediately after ingestion of the fish. A variety of neurological symptoms may be present, and death may ensue in less than 15 minutes. The relationship of clupeotoxism to ciguatera fish poisoning, if any, has not been determined since nothing is known concerning the nature of clupeotoxin.

Prevention: One cannot detect a poisonous fish by its appearance. Moreover, there is no known simple chemical test to detect the poison. The most reliable methods involve the preparation of tissue extracts which are injected intraperitoneally into mice, or feeding samples of the viscera and flesh to cats or dogs, and observing the animal for the developments of toxic symptoms. The viscera—liver and intestines—of tropical marine fishes should *never* be eaten. Also, the roe of most marine fishes is potentially dangerous, and in some cases may produce rapid death. Fishes which are unusually large for their size should be eaten with caution. This is particularly true for barracuda (*Sphyraena*), jacks (*Caranx*), and grouper (*Epinephelus*) during their reproductive

Figure 99. A. **Squaretail,** *Tetrogonurus cuvieri* (Risso). Length 10 inches (25 centimeters). ▸ B. **Surgeonfish,** *Acanthurus glaucopareius* (Cuvier). Length 8 inches (20 centimeters). C. **Surmullet,Goatfish,** *Parupeneus chryserydros* (Lacépède). Length 12 inches (30 centimeters). D. **Surmullet,Goatfish,** *Upeneus arge* (Jordan and Evermann). Length 12 inches (30 centimeters).

A
B

seasons. Do not eat moray eels. Even some of the "safe" species may at times be violently poisonous.

If one is living under survival conditions, and questionable fishes must be eaten, it is advisable to cut the fish into thin fillets and to soak them in several changes of water—fresh or salt—for at least 30 minutes. Do not use the rinse water for cooking purposes. This serves to leach out the poison which is somewhat water soluble. If a questionable species is cooked by boiling, the water should always be discarded. It must be emphasized that ordinary cooking procedures do not destroy or significantly weaken the poison. The advice of native people on eating tropical marine fishes is sometimes conflicting and erroneous, particularly if they have not lived within a particular region over a period of time. Nevertheless, one should always check with the local natives as to the edibility of fish products in any tropical island area. Keep in mind that an edible fish in one region may be deadly poisonous in another region.

SCOMBROID POISONING

This form of poisoning involves the scombroid fishes which includes tuna, mackerel, skipjack, and related species. The fishes included within this category are usually edible and valuable commercial species. Scombroid poisoning is caused by improper preservation which results in certain bacteria on the histidine in the muscle of the fish which in turn converts the histidine to saurine, a histaminelike substance. This is the only form of ichthyosarcotoxism in which bacteria play an active role in the production of the poison within the body of the fish. However, the poison is not a bacterial endotoxin.

A list of the species will not be given since any of the tuna, skipjack, bonito, mackerel, etc., may be involved. Representatives of these fishes are world-wide in their distribution.

Medical Aspects: Symptoms usually develop within a few minutes after ingestion of the fish. Frequently, toxic fish can be detected immediately upon tasting because of a "sharp or peppery" taste. The symptoms of scombroid poisoning resemble those of a histamine intoxication, including intense headaches, dizziness, throbbing of the large blood vessels of the neck, epigastric pain, flushing of the face, generalized erythema, urticarial eruptions, severe itching, bronchospasm, burning of the throat, cardiac palpitation, nausea, vomiting, diarrhea, abdominal pain, thirst, inability to swallow, suffocation and severe

A

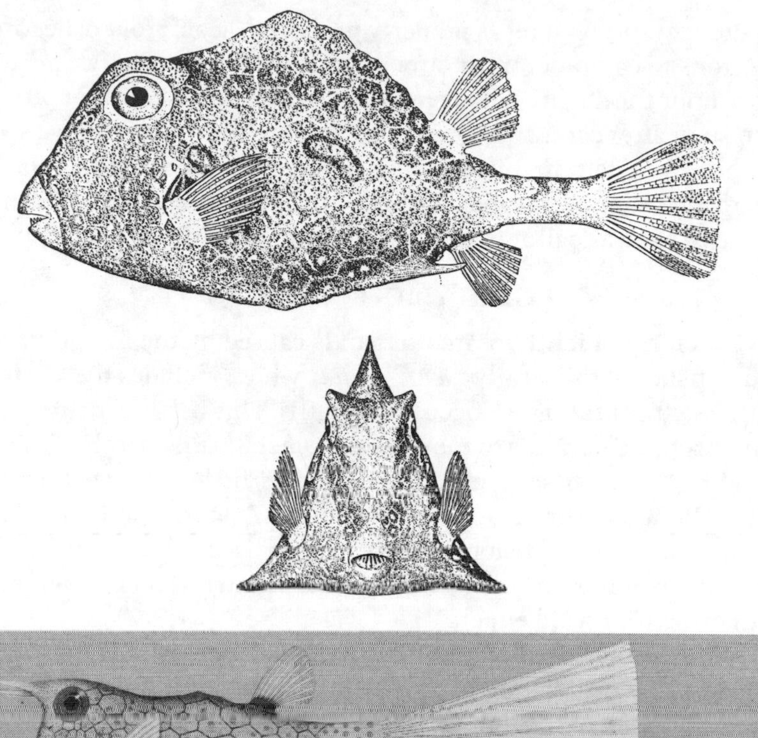

B

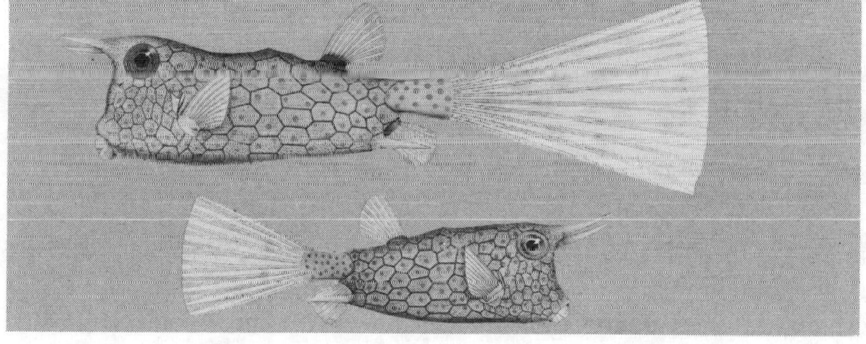

Figure 100. A. **Trunkfish,** *Lactophrys trigonus* (Linnaeus). Length 12 inches (30 centimeters). B. **Trunkfish,** *Lactoria cornutus* (Linnaeus).

respiratory distress. There is a danger of shock, and deaths have been reported. However, the victim usually recovers within a period of a day or two. Scombrotoxism is believed to be the most common form of ichthyosarcotoxism on a world-wide basis.

Treatment: In addition to such routine procedures as evacuation of the stomach and catharsis, the use of any of the ordinary antihistaminic drugs will be found to be effective.

Prevention: As long as properly preserved, scombroid fishes are not dangerous to eat under most circumstances. Commercially canned fish are without the slightest danger. Scombroids should be either promptly eaten soon after capture or preserved, by canning or freezing, as soon as possible. Fish left in the sun for longer than two hours should be discarded. Examine the fish before eating; if there is any evidence of staleness, such as pallor of the gills or an off-odor, discard the fish.

GEMPYLID POISONING

This form of ichthyosarcotoxism is caused by the ingestion of the flesh of fishes of the family *Gempylidae*, which includes the **Castor Oil Fish,** *Ruvettus pretiosus* Cocco (Figure 101). This fish is usually found in deep water, 400 fathoms or more, and is generally taken by hook and line at night. Although seasonal in occurrence, little is known about its habits. Ranging throughout the tropical Atlantic and Indo-Pacific Oceans, the castor oil fish is properly named because of its oily flesh and bones. Ingestion of the oil produces a diarrhea which is without cramping or other untoward effects.

No treatment is required. Many natives esteem this fish for food because of its mild purgative effect. Persons eating the fish should merely be aware of this purgative effect. The diarrhea can be stopped by not eating the fish.

HALLUCINOGENIC FISH POISONING

A number of different species of tropical reef fishes have been known to produce hallucinations after they have been ingested. Two fishes which appear to be more commonly involved than any of the others include the **Sea Chub,** *Kyphosus cinerascens* (Forskål) (Figure 101) and the **Goat Fish,** *Upeneus arge* Jordan and Evermann (Figure 99). Both species are found in the Indo-Pacific region. The type of poisoning caused by ingesting these fishes is sporadic and unpredictable in its occurrence.

Medical Aspects: The poison affects primarily the central nervous system. The symptoms usually develop within a period of two hours, may persist for 24 hours or more, and include dizziness, loss of equilibrium, lack of motor coordination, abdominal cramps, nausea, vomiting, hallucinations, and mental depression. A tight constriction of the chest or the victim stating, "It feels like someone is sitting on my chest," are typical complaints. The victim frequently feels that he may be going to die. Actually, this is a mild form of fish poisoning, and no fatalities have been reported. Nothing is known about the nature of the poison.

Treatment: Symptomatic, but usually no treatment is required.

Prevention: Check the edibility of the fish with a local native.

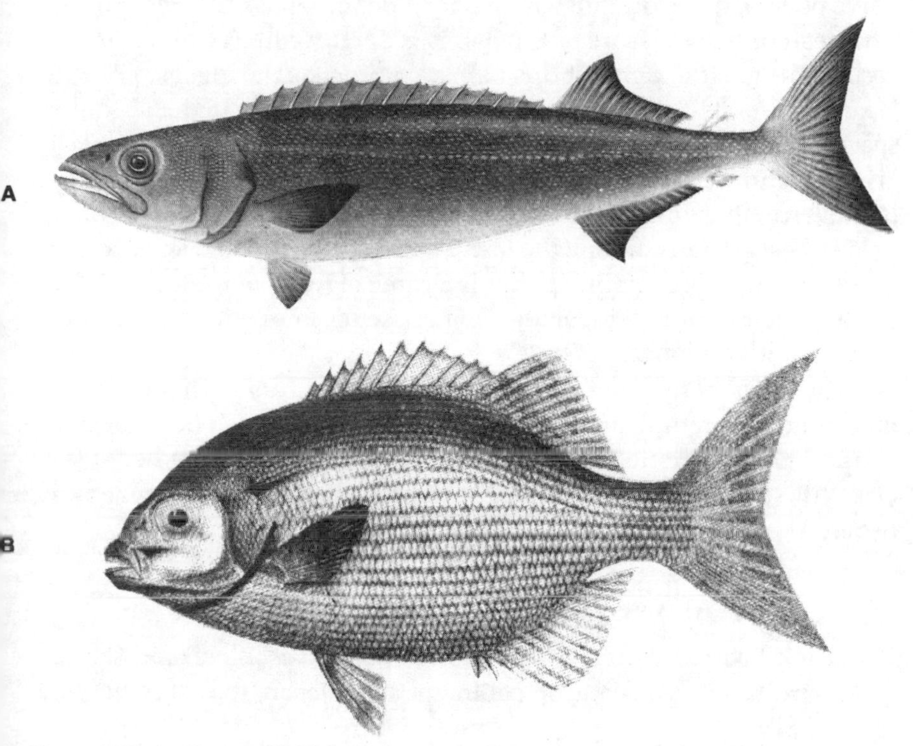

Figure 101. A. **Castor Oil Fish,** *Ruvettus pretiosus* Cocco. Length 4.1 feet (1.4 meters) (From Hiyama) B. **Sea Chub,** *Kyphosus cinerascens* (Forskål). Length 20 inches (50 centimeters). (From Marshall)

PUFFER POISONING

This form of ichthyosarcotoxism is one of the most violent forms of poisoning known. It is caused by a group of fishes of the Order *Tetraodontoidea,* which includes the puffers, porcupinefishes, and the ocean sunfishes. Puffers, which are referred to by a variety of names such as the globefish, blowfish, balloonfish, toadfish, toado, swellfish, fugu, botete, fahaka, tinga, etc., are all characterized by their remarkable ability to inflate themselves by gulping in large quantities of air or water. They make considerable noise during inflation by grinding their heavy jaw teeth together. Some puffers can, and do, inflict nasty bites. Puffers have a distinctive offensive odor, which is particularly noticeable when they are dressed. Of all marine creatures puffers are among the most poisonous. The liver, gonads, intestines, and skin usually contain a powerful nerve poison, tetrodotoxin, which may produce rapid and violent death. The flesh or musculature of the fish is generally edible. Strange to say, despite the great toxicity of this fish, it commands the highest prices in Japan as a food. Called fugu in Japan, puffers are prepared and sold in special restaurants where highly trained fugu cooks are used. The fugu is given careful treatment by these chefs to eliminate the danger of eating it. Nevertheless, it is still the number one cause of fatal food poisoning in Japan—especially among the lower classes, who fail to take necessary precautions. At best, eating puffer is a game of Russian roulette. Unless the services of a professional fugu connoisseur can be obtained, it is best to leave puffers alone.

Although puffers are most numerous in the tropics, many species do extend into temperate zones (Figure 102). Puffers can be recognized by their characteristic shape and large teeth. The following list of species will serve to represent some of the more poisonous tetraodontoid fishes.

REPRESENTATIVE SPECIES OF POISONOUS TETRAODONTOID FISHES

Black-Spotted Puffer, *Arothron nigropunctatus* (Bloch and Schneider) (Plate 12). Polynesia, tropical Indo-Pacific Japan, to east Africa and the Red Sea.

Gulf Puffer, *Sphaeroides annulatus* (Jenyns) (Plate 12). Ranges from California to Peru, and the Galapagos Islands.

Maki-Maki, or **Deadly Death Puffer,** *Arothron hispidus* (Linnaeus) (Plate 12). Ranges from Panama throughout the tropical Pacific, Japan, to South Africa and the Red Sea.

Porcupine Fish, *Diodon hystrix* Linnaeus (Plate 13). Circumtropical in its distribution, occasionally entering temperate areas.

White-Spotted Puffer, *Arothron meleagris* (Lacépède) (Plate 12). Ranges from the west coast of Central America to Indonesia.

Medical Aspects: Puffer poisoning is characterized by paresthesias, a tingling sensation of the lips and tongue which gradually spreads to include the extremities and later develops into severe numbness involving the entire body. Gastrointestinal disturbances may or may not be present. Respiratory distress is a prominent part of the clinical picture, and the victim eventually becomes intensely cyanotic. Petechial hemorrhages, blistering and severe scaling of the skin may develop. Ataxia,

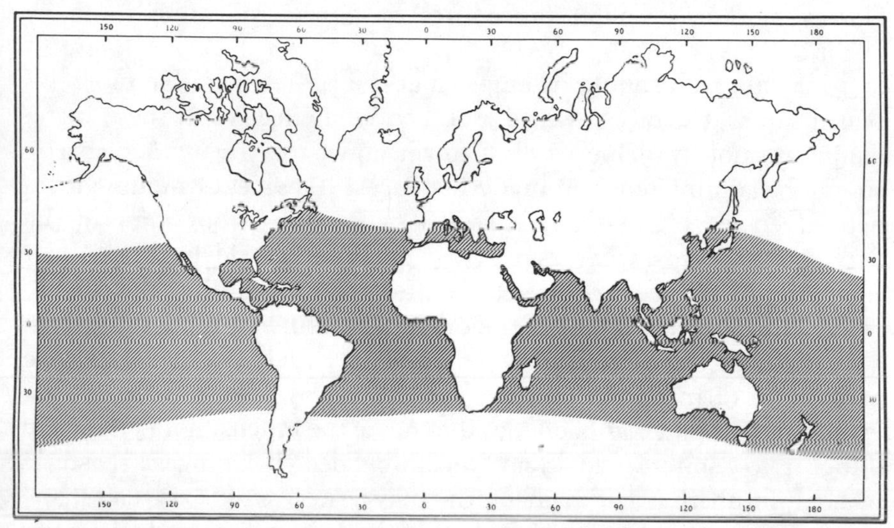

Figure 102. Map showing the geographical distribution of puffers and pufferlike fishes.

aphonia, dysphagia, muscular twitchings, tremors, incoordination, paralysis and convulsions are frequently present. The victim may become comatose, but in most instances remains conscious until shortly before death. Treatment is symptomatic. The case fatality rate is about 60 percent. If death occurs, it usually takes place within the first 24 hours.

Pharmacological studies have shown that the primary action of tetrodotoxin is on the nervous system, producing both central and peripheral effects. Relatively low doses of the poison will readily inhibit neuromuscular function. Major effects include respiratory failure and hypotension. It is believed that puffer fish poison has a direct action on respiratory centers; there is also some evidence that tetrodotoxin depresses the vasomotor center.

Treatment: See treatment of fish poisoning. There is no specific treatment or antidote for puffer poisoning.

Prevention: Learn to recognize the puffer and leave it alone. It may make an excellent poisonous bait for stray cats, but a poor food for humans.

GENERAL TREATMENT FOR FISH POISONING

With the exception of scombroid poisoning in which the patient should be administered antihistaminic drugs, there is no specific treatment. However, a few general procedures have been of value in many instances.

The stomach should be emptied at the earliest possible moment. Warm salt water, or egg white, will be found effective. If these ingredients are not available, stick a finger down the throat. A cathartic should be administered. In many instances, 10 percent calcium gluconate given intravenously has given prompt relief from some of the nervous symptoms, whereas in others it has not. Paraldehyde and ether inhalations have been reported to be effective in controlling the convulsions. Nikethamide or one of the other respiratory stimulants is advisable in cases of respiratory depression. In patients where excessive production of mucus is present, aspiration and constant turning are essential. Atropine has been found to make the mucus more viscid and difficult to aspirate and is not recommended. If laryngeal spasm is present, intubation and tracheotomy may be necessary. Oxygen inhalation and intravenous administration of fluids supplemented with vitamins given parenterally are usually beneficial. If the pain is severe, opiates will be required. Morphine is the drug of choice when given in small, divided doses. Cool showers have been found to be effective in relieving the severe itching. It should be kept in mind that in rare instances scombroid poisoning may be combined with other types of fish poisoning. Fluids given to patients suffering from disturbances of temperature sensation should be slightly warm, or at room temperature. Vitamin B complex supplements are advisable.

POISONOUS MARINE TURTLES

Poisoning from marine turtles is one of the lesser known types of intoxications produced by marine organisms. The cases that have been reported are sufficiently severe to be impressive. As in the case of fishes,

Figure 103. A. **Green Sea Turtle,** *Chelonia mydas* (Linnaeus). Length of carapace 4 feet (1.2 meters). B. **Hawksbill Turtle,** *Eretomochelys imbricata* (Linnaeus). Length of carapace 34 inches (85 centimeters). C. **Leatherback Turtle,** *Dermochelys coriacea* (Linnaeus). Length of carapace 4 feet (1.2 meters).

A
B
C

Figure 104 A. **Bearded Seal,** *Erignathus barbatus* (Erxleben). Length 9 feet (2.7 meters). B. **Australian Sea Lion,** *Neophoca cinerea* (Péron). Length 8 feet (2.4 meters).

most of these species are commonly eaten with impunity. For some unknown reason, certain species of marine turtles in the vicinity of the Philippine Islands, Sri Lanka, and Indonesia, under certain circumstances, may become extremely poisonous to eat.

SPECIES OF MARINE TURTLES REPORTED AS POISONOUS TO EAT

Green Sea Turtle, *Chelonia mydas* (Linnaeus) (Figure 103). Inhabits all tropical and subtropical seas.

Hawksbill Turtle, *Eretomochelys imbricata* (Linnaeus) (Figure 103). Inhabits all tropical and subtropical seas.

Leatherback Turtle, *Dermochelys coriacea* (Linnaeus) (Figure 103). Largely circumtropical, but occasionally taken in temperate waters.

Medical Aspects: Symptoms generally develop within a few hours to several days after ingestion of the flesh. The initial symptoms are usually nausea, vomiting, diarrhea, severe upper abdominal pain, dizziness, dry burning sensation of the lips, tongue, lining of the mouth and throat. Swallowing becomes very difficult, and excessive salivation is pronounced. The disturbances of the mouth may take several days to develop but become progressively severe as time goes on. The breath becomes very foul. The tongue develops a white coating and may become covered with multiple pinhead-sized, reddened papules which may later break down into ulcers. If the victim has been severely poisoned, he tends to become very sleepy and is difficult to keep awake. If this symptom develops, it is usually a bad sign as death soon follows. About 44 percent of the victims poisoned by marine turtles die. Death is believed to be due to liver and kidney damage.

Treatment: There is no specific treatment. Some of the recommendations presented in fish poisoning are pertinent here.

Prevention: Marine turtles in the tropical Indo-Pacific region should be eaten with caution. If in doubt, a check with local native groups will determine if they are safe to eat in that locality. Turtle liver is especially dangerous to eat.

MARINE MAMMALS

Several species of marine mammals have been incriminated in human intoxications. These include:

Australian Sea Lion, *Neophoca cinerea* (Péron) (Figure 104). This sea lion is confined to the coast of South Australia. The flesh of this animal is said to be toxic, but there is very little supporting data upon which to base any conclusions.

Bearded Seal, *Erignathus barbatus* (Erxleben) (Figure 104). This species is circumboreal in its distribution, living at the edge of ice. Its

Figure 105. **Polar Bear,** *Thalarctos maritimus* Phipps. Attains a length of 8 feet (2.5 meters) and a weight of 1500 pounds (750 kilograms). Ingestion of the liver and kidneys of the polar bear may result in a fatal intoxication due to hypervitaminosis A.

liver is said to contain a high concentration of Vitamin A, and may be as toxic as polar bear liver.

Polar Bear, *Thalarctos maritimus* Phipps (Figure 105). Ranges throughout the Arctic regions of the world. Numerous poisonings have resulted from eating the liver and kidneys of polar bears. The predominant symptoms are intense throbbing or dull frontal headaches, nausea, severe scaling of the skin, vomiting, diarrhea, abdominal pain, dizziness, drowsiness, irritability, collapse, photophobia, and convulsions. Fatalities are rare. Polar bear poisoning consists primarily of central nervous system manifestations primarily due to an abrupt and marked elevation of spinal fluid pressure.

Polar bear poisoning is believed to be related to an excessive intake of Vitamin A which is known to be present in the liver and kidneys of these bears. Symptoms develop in adults with the ingestion of about one million international units of Vitamin A. Seven million international units may be fatal.

It has been suggested that other toxic substances in addition to Vitamin A may also be present, but this has not been determined with any degree of certainty.

WHALES AND DOLPHINS

Several species of whales and dolpins are reported to have poisonous flesh, but data is too meager at this time to permit any definitive statements.

Treatment: Emetics and laxatives promptly administered are sometimes useful in relieving symptoms. The clinical symptoms gradually disappear after ingestion of the toxic material has been discontinued.

Prevention: There is no reliable method of detecting toxic marine mammal poisons by mere visual examination. The age of the animal has no bearing on its edibility; young animals may cause poisoning as well as older ones.

VII

OTHER DANGEROUS AQUATIC ANIMALS

Included in this brief chapter are several groups of noxious aquatic organisms which are capable of inflicting injuries to humans. Some of these organisms are found only in freshwater. They are included in the present volume because they may be encountered in a water environment likely to be visited by adventurous divers. Furthermore, a discussion about noxious aquatic animals would not be complete without some reference to them.

MARINE ANIMALS THAT SHOCK

Electric fishes constitute a relatively minor health hazard, but, nevertheless, they are worthy of serious consideration. There are several different groups of fishes that possess electric organs: catfishes (*Malopterurus*), star-gazers (*Astroscopus*), electric eels (*Electrophorus*), and a number of genera of electric rays (*Torpedo, Narcine, Hypnarce, Hypnos, Discopyge,* etc.). Electric eels and catfishes are freshwater inhabitants, whereas star-gazers and rays are marine. The most important marine members are the electric rays, representatives of which are found in all temperate and tropical oceans. Electric rays such as *Narcine brasiliensis* (Olfers) (Figure 106) are sluggish, feeble swimmers, spending most of their time lying on the bottom partially buried in the mud or sand, generally preferring shallow depths.

The electric organs, which constitute about one-sixth of the total body weight of the ray, are situated one on either side of the anterior part of the disc between the anterior extension of the pectoral fin and the head, extending from about the level of the eye backward past the gill region. Usually, outlines of the organs are externally visible on both the ventral and dorsal sides. The organs are comprised of columnar prism-like structures separated by loose connective tissue, forming a network similar to the cells of a honeycomb. The columns vary in number according to the species of ray. The ventral side of the ray is electrically negative whereas the dorsal side is positive. The production of an elec-

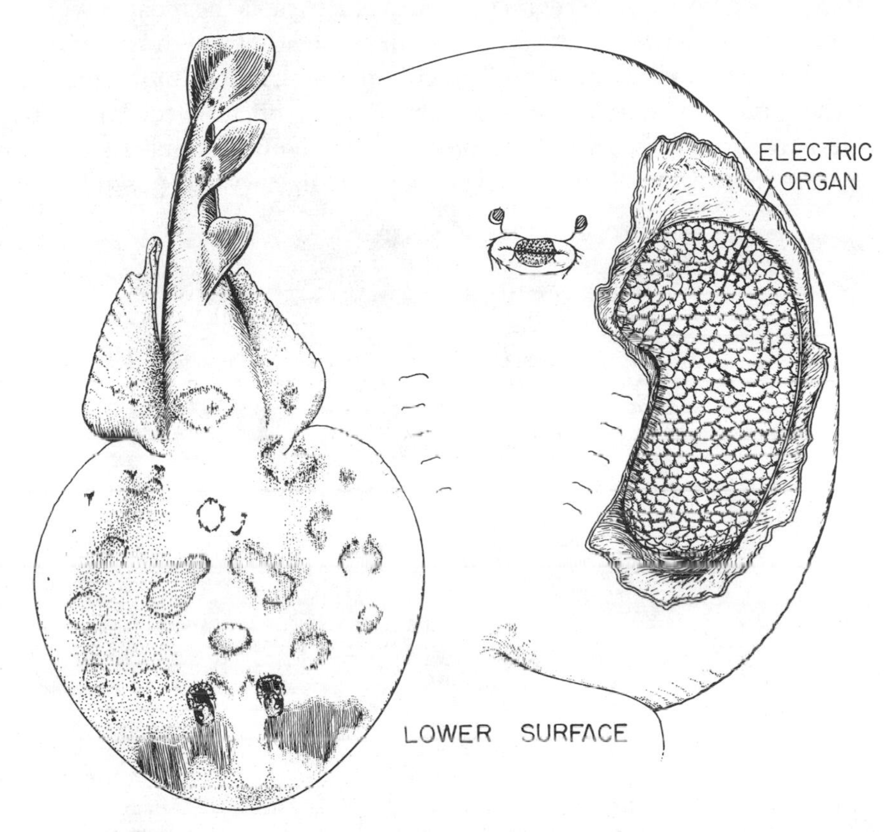

Figure 106. A. **Lesser Electric Ray,** *Narcine brasiliensis* (Olfers). Length 11 inches (45 centimeters). B. Showing the lower surface of the electric ray, *Narcine brasiliensis.* The skin is removed to reveal the electric organ. Contact with the skin surface of the ray can produce a shocking experience! (After Bigelow and Schroeder)

trical discharge or shock is believed to be a simple reflex action—the result of tactile stimulation. A ray can deliver a successive series of discharges, but becomes progressively weaker, until finally exhausted. After a period of time, the ray recuperates and again is able to produce electrical discharges. The voltage delivered varies with the individual species, but is said to range from 8 to 220 volts. Completion of the circuit by contacting the ray at two points is not necessary if the ray is in the water. Contact with a large ray may result in a shock sufficient to knock over and temporarily disable a human being. Recovery is usually uneventful.

The electric eel *Electrophorus electricus* (Linnaeus) (Figure 107), found in the freshwater streams of South America, is the most powerful of the electric fishes. *Electrophorus* is an air breather and must come to the surface periodically for gulps of air, drowning if kept submerged all of the time. When at rest *Electrophorus* gives off no electricity, but when cruising about will emit a discharge of about 50 volts at a rate of about 50 per second. Developing cataracts at an early age, possibly due

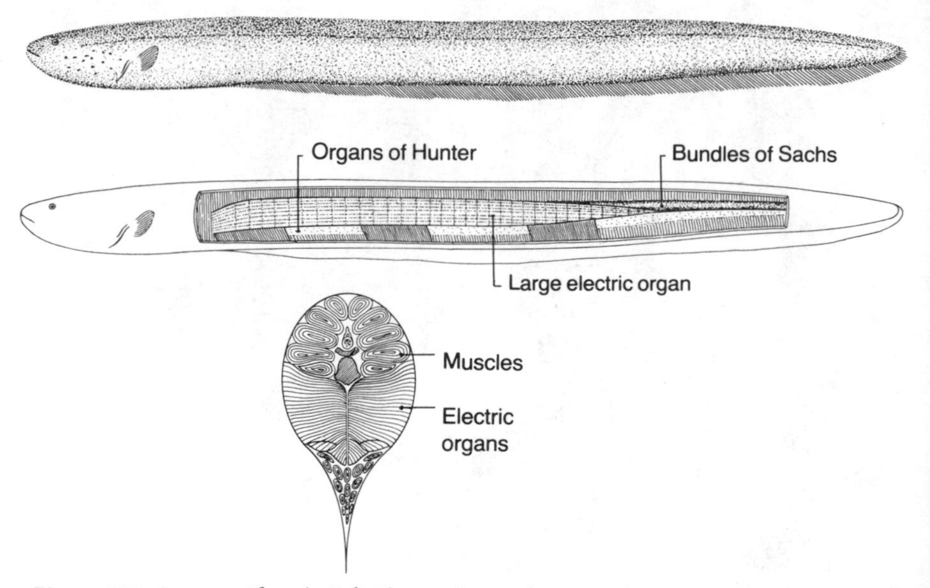

Figure 107. **Amazon Electric Eel,** *Electrophorus electricus* (Linnaeus). Length 4 feet (1.2 meters). An electric eel can deliver a painful shock sufficient to stun a man or a horse. Cutaway drawings show the anatomical distribution of the electric organs. (After Coates)

to their electrical discharges, electric eels use these discharges as a sonic device to detect food, to ward off enemies, and to act as a general guidance system. The discharge of an electric eel varies from about 370 to 550 volts with an average of about 40 watts, sufficient to stun a man or a horse. The discharge lasts only two one-thousandths of a second, but the eel can send out four hundred or more per second. An eel can give out a steady series of discharges for 20 minutes, rest for 5 minutes, and then continue the shocking process again, making this eel, no doubt, the world's most efficient battery. In ancient times the shocking process of electric fishes was used in the treatment of rheumatism. Figure 107 shows the anatomical arrangement of the electric organs in *Electrophorus*.

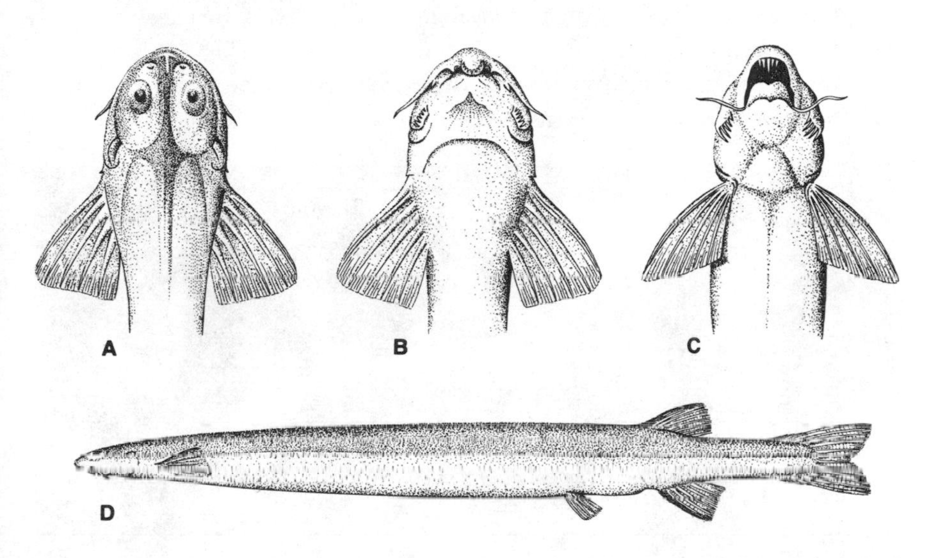

Figure 108. Drawings of the parasitic candiru catfish of the Amazon. A. Dorsal view of head of *Urinophilus erythrurus* Eigenmann. B. Ventral view of the head showing opercular spines. C. Ventral view of head of *Vandellia cirrhosa* Cuvier and Valenciennes. Note teeth of upper jaw and gill covers. D. *Urinophilus erythrurus* Eigenmann. (After Gudger)

HUMAN PARASITIC CATFISH

Stories concerning catfishes that penetrate the human urethra have issued from the Amazon jungle for centuries. It is now known that these catfishes are members of the family *Pygidiidae,* and one genus has been appropriately designated as *Urinophilus,* or a "lover of urine." Natives refer to these fish as the "candirú" or "canero." Documented reports have cited instances in which the candirú is known to have penetrated the urethra, vagina and rectum of humans and other animals. Most of the candirú capable of urinary misconduct are either juvenile or small specimens. They are generally elongated or eellike in form (Figure 108), cylindrical to terate in cross section, have a rounded depressed head, and an inferiorly placed mouth. The body of these fishes are scaleless, slimy, having a transparent or flesh to brownish color. The dangerous candirú are about an eighth to 2 inches (3 millimeters - 50 millimeters) in length. The candirú is a formidable parasite to have clinging to one's urethra because of its fanglike erectile teeth in front of the mouth, and the opercular spines situated on either gill cover, and on the head (Figure 109). Natives wading in streams inhabited by the candirú wear a variety of protective devices hoping to barricade any intrusion by these devilish beasts (Figure 110).

Once the candirú invades the urethra, vagina, or rectum they usually have to be surgically removed.

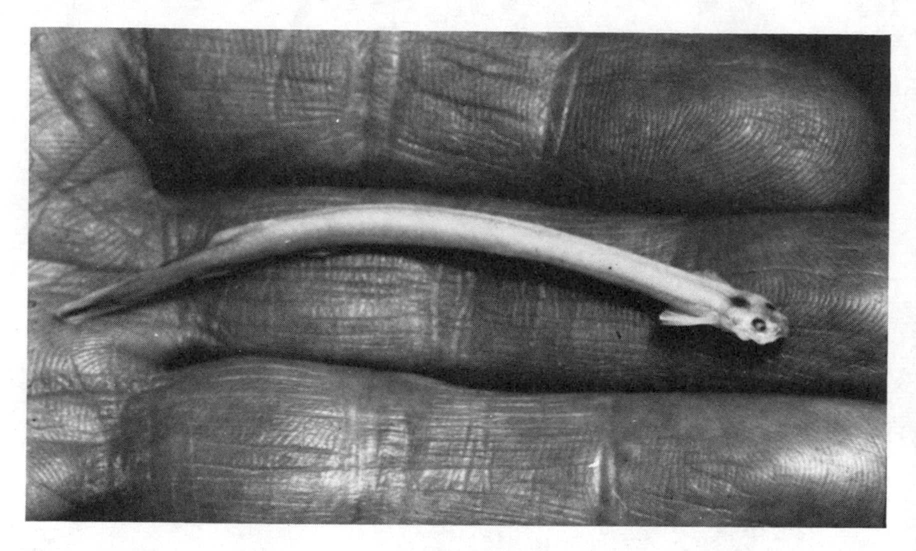

Figure 109. Photograph of *Urinophilus erythrurus*. Because of its shape and small size, this catfish is capable of invading the urethra and other orifices of the body.

SCHISTOSOMAL DERMATITIS

This is a type of dermatitis due to the penetration of the skin by one of the larval stages of various species of flukes. The infestation may occur in freshwater, brackish, or salt water.

The flukes are members of the phylum *Platyhelminthes,* which includes the class *Turbellaria,* the planarians; the class *Cestoda,* the tapeworms; and the class *Trematoda,* the flukes. Under normal circumstances, the definitive host of many fluke species are parasites of a wide variety of vertebrates, including human beings in some instances. In the case of marine schistosomal dermatitis the definitive host is a marine bird or some other marine vertebrate.

The eggs of the fluke are dropped into the water and hatch into an early larval stage which eventually invades a suitable marine snail which serves as an intermediate host.

Medical Aspects: The larvae of the fluke, known as *cercariae*, are released from the snail, whereupon they seek out a suitable host for their further development. It is at this point that they become involved

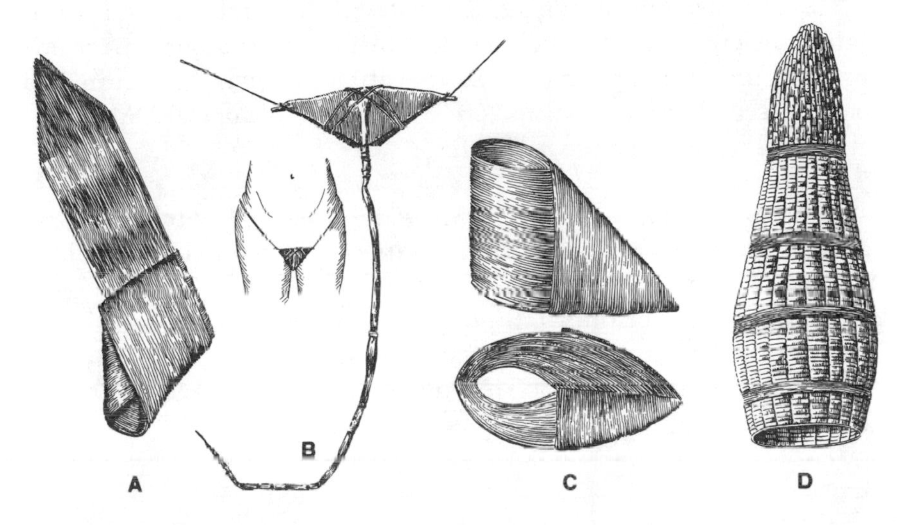

Figure 110. Protective devices worn by natives to protect themselves from the parasitic candirú, and from schistosomal infections. A. Penisstulp worn by men of the Baoró tribe, Xingu River, Amazonia, Brazil. B. Wuri, or pudendal covering worn by Bakairi women, upper Xingu River. C. Renisstulp worn by men of the Kanapó and Tapairape tribes along the Araguana River, Amazonia. D. Protective device worn by the Zulus of Rhodesia to prevent penetration of the penis by the trematode parasite *Schistosoma*. (After Gudger)

with human beings. Upon penetration of human skin the cercariae are capable of producing a severe skin reaction. At the site of entry an intense local foreign body reaction is produced which results in the death of the larval fluke. The death of the larvae causes a reddened papular eruption with intense itching. Repeated infestations can cause severe hypersensitivity reactions.

The first symptoms of cercarial infestation usually occur as a prickling sensation either in the water, or soon after leaving it, possibly due to the mechanical irritation of the penetration of the skin. The sensation soon disappears, but a small red spot remains where the cercaria penetrated the skin. If the person is hypersensitive to these worms, the red spot becomes papular, inflammed and swollen, accompanied by intense itching. Due to scratching or rubbing, the lesions can become severely infected. In most cases after several days the spots begin to fade, and in about a week only brown spots remain.

Schistosomal dermatitis has been reported in New South Wales and Northern Territory, Australia, Tasmania, Hawaii, and Florida.

Treatment: Antihistamines are useful in controlling the inflammatory reaction and the itching. Antihistamine and steroid creams provide prompt symptomatic relief. Antibiotics may be required to control secondary bacterial infections. The use of topical DMSO (dimethylsulfoxide) may also be helpful.

Prevention: Persons bathing in waters infested with flukes are advised to use protective clothing—either a heavy nylon or rubber diving suit. Dimethylpthallate solution (25 percent in a cream base) rubbed on the skin before swimming in an infested area is reported to be helpful in minimizing the cercarial infestation. Brisk rubbing with a rough towel immediately upon leaving the water is also helpful.

SELECTED BIBLIOGRAPHY

Baldridge, H.D. 1975. *Shark attack.* New York: Berkley Publishing Corp., 263 p.

Baslow, M.H. 1969. *Marine pharmacology.* Baltimore: Williams & Wilkins Co., 286 p.

Bigelow, H.F. and Schroeder W.C. 1948. Fishes of the western North Atlantic. Sharks. *Sears Found. Mar. Res.*, Mem. No. 1, Pt. 1:59-546, figs. 7-106.

Bolin, R.L. 1954. Report on a fatal attack by a shark. *Pac. Sci.* 8:105-108, 2 figs.

Bottard, A. 1889. *Les poissons venimeux.* Paris: Octave Doin., 198 p.

Brown, T.W. 1973. *Sharks—The search for a repellent.* Sydney, Australia: Anchor Books Pty. Ltd., 134 p.

Budker, P. 1947. *La vie des requins.* Villeneuve-Saint-Georges: Gallimard., 227 p., 22 pls., 40 figs.

———. 1971. *The life of sharks.* London: The Trinity Press., 222 p.

Cleland, J.B. and Southcott, R.V. 1965. *Injuries to man from marine invertebrates in the Australian region.* Canberra, Australia: A.J. Arthur, Commonwealth Government Printer, 282 p.

Coates, C.W. 1947. The kick of an electric eel. *Atlantic Monthly.* 180(4):75-79.

———. 1950. Electric fishes. *Elec. Eng.* pp. 1-3 (Jan.).

Coates, C.W.; Cox, R.T.; and Granath, L.P. 1937. The electric discharge of the electric eel, *Electrophorus electricus* (Linnaeus). *Zoologica, New York.* 22(1):1-32, 6 figs., 2 pls., 2 tabs.

Coppleson, V.M. 1950. A review of shark attacks in Australian waters since 1919. *Med. Journ. Australia,* 2:680-687, 7 figs.

Cousteau, J. and Cousteau, P. 1970. *The shark: splendid savage of the sea. The undersea discoveries of Jacques-Yves Cousteau.* New York: A & W Pubs., Inc. 273 p.

Davies, D.H. 1964. *About sharks and shark attacks.* Durban, South Africa: Brown Davis and Platt Ltd., 237 p.

DeSylva, Donald P. 1963. Systematics & life history of the great barracuda. (*Studies in Tropical Oceanography* Ser. No. 1) U Miami Marine.

Dunson, W.A. 1975. *The biology of sea snakes.* Baltimore: University Park Press, 530 p.

Edmonds, C. 1975. *Dangerous marine animals of the Indo-Pacific Region.* Newport, Australia: Wedneil Publications., 235 p.

Evans, H.M. 1943. *Sting-fish and seafarer.* London: Faber and Faber, Ltd., 180 p., 7 pls., 30 figs.

Fisher, A. A. 1978. *Atlas of aquatic dermatology.* New York: Grune & Stratton, 113 p.

Gilbert, P.W. and members of the Shark Research Panel of the American Institute of Biological Sciences., eds. 1963.*Sharks and survival.* Boston: D.C. Heath and Company, 578 p.

Gilbert, P.W.; Mathewson, R.F.; and Rall, D.P., eds. *Sharks, snakes, and rays.* Baltimore: The Johns Hopkins Press., 624 p.

Gudger, E.W. 1918. *Sphyraena barracuda;* its morphology, habits, and history. *Pub. Carnegie Inst.* Washington, 12(252):53-108, 2 figs., 7 pls.

———. 1930. On the alleged penetration of the human urethra by an Amazonian catfish called candiru. *Amer. Jour. Surg.*, 8(1):170-188, (13 text-figs.) 8(2):443-457 (4 text-figs.).

Halstead, B.W. 1965. *Poisonous and venomous marine animals of the world. Vol. one—Invertebrates.* Washington, D.C.: Government Printing Office, 994 p.

———. 1967. *Poisonous and venomous marine animals of the world. Vol. two—Vertebrates.* Washington, D.C. Government Printing Office, 1070 p.

———. 1970. *Poisonous and venomous marine animals of the world. Vol. two—Vertebrates (cont'd.)* Washington, D.C.: Government Printing Office, 1006 p.

———. 1978. *Poisonous and venomous marine animals of the world.* Princeton: Darwin Press, Inc., 1043 p.

———. 1980. *Field guide to poisonous and venomous marine animals of the Indo-Pacific.* Princeton: Darwin Press, Inc., in press.

Hiyama, Y. 1950. Poisonous fishes of the South Seas. *Spec. Sci. Rept., U.S. Fish and Wildlife Serv.*, No. 25: 188 p.

Kaiser, E. and Michl, H. 1958. Die biochemie der tierischen gifte. Einzeldarstellungen aus dem gesamtgebiet der biochemie. *Neue Folge.* II Bd. Vienna: Franz Deuticke, 258 p., 23 abb., 57, Tb.

Lineaweaver, III, T.H. and Backus, R.H. 1969. *The natural history of sharks.* Philadelphia: Lippincott, J.B. Co., 256 p.

National Academy of Sciences. 1973. *Toxicants occurring naturally in foods.* 2nd ed. Washington, D.C.: Committee on Food Protection, Food and Nutrition Board, National Research Council., 624 p.

Pawlowsky, E.N. 1927. *Gifttiere und ihre Giftigkeit.* Jena: Gustav Fischer., 516 p., 176 figs.

Phisalix, M. 1922. *Animaux venimeux et venins.* Paris: Masson et Cie. 1:656 pp.; 2:864 p.; 17 pls., 521 figs.

Russell, F.E. 1953. Stingray injuries: a review and discussion of their treatment. *Amer. Journ. Med. Sci.*, 226:611-622, 3 figs.

———. 1965. Marine toxins and venomous and poisonous marine animals. Reprinted from Russell, Frederick S., ed., *Advances in marine biology*, Vol. 3. New York: Academia Press, Inc., 255-384 p.

Whitley, G.P. 1940. *The fishes of Australia. Part 1. The sharks, rays, devil-fish, and other primitive fishes of Australia and New Zealand.* New South Wales, Australia: Roy. Zol. Soc., 280 p., 303 figs.

INDEX

ABOUT THE AUTHOR

Bruce W. Halstead, one of the world's leading authorities on biotoxicology, is director of the World Life Research Center in Colton, California, and medical director of the Halstead Preventive Medical Clinic in Loma Linda.

A native of southern California, he graduated from San Francisco City College, the University of California, Berkeley, and in 1948 received an M.D. from the Loma Linda University School of Medicine.

Dr. Halstead served in the Enlisted Reserve Corps, U.S. Army, the U.S. Public Health Service, and in the U.S. Navy as a commander in the Medical Corps.

In his active role as scientist, consultant, lecturer, and author, Dr. Halstead has traveled to more than 150 nations and has contributed over 200 articles to scientific publications and diving journals. His research has been published in six languages.